Microscopic Wonders

Abraham Rosas-Arellano
Carmen Reyes Luna
Fabiola García-Zamorategui
Ricardo Piña-Muñoz
Yazmín Ramiro-Cortés
Gerardo Rodrigo Perera-Murcia
Alfonso Cárabez-Trejo

Microscopic Wonders

The Science of Seeing the Invisible

Abraham Rosas-Arellano (iD)
Imaging Unit
National Autonomous University of Mexico
Institute of Cellular Physiology
CDMX, Mexico City, Mexico

Fabiola García-Zamorategui (iD)
Yinfayang Tattoo Studio
Mexico City, Mexico

Yazmín Ramiro-Cortés
Department of Neurodevelopment and
Physiology
National Autonomous University of Mexico
CDMX, Mexico City, Mexico

Alfonso Cárabez-Trejo
Department of Developmental
Neurobiology and Neurophysiology
National Autonomous University of Mexico
Institute of Neurobiology
Juriquilla, Queretaro, Mexico

Carmen Reyes Luna
Metsera Inc.
London, UK

Ricardo Piña-Muñoz (iD)
Department of Biology
Metropolitan University of Educational
Sciences
Santiago, Chile

Gerardo Rodrigo Perera-Murcia
Department of Neurodevelopment and
Physiology
National Autonomous University of Mexico
Institute of Cellular Physiology
CDMX, Mexico City, Mexico

ISBN 978-3-031-92558-0   ISBN 978-3-031-92559-7   (eBook)
https://doi.org/10.1007/978-3-031-92559-7

© The Editor(s) (if applicable) and The Author(s), under exclusive license to Springer Nature
Switzerland AG 2025

This work is subject to copyright. All rights are solely and exclusively licensed by the Publisher, whether
the whole or part of the material is concerned, specifically the rights of translation, reprinting, reuse of
illustrations, recitation, broadcasting, reproduction on microfilms or in any other physical way, and transmission or information storage and retrieval, electronic adaptation, computer software, or by similar or
dissimilar methodology now known or hereafter developed.
The use of general descriptive names, registered names, trademarks, service marks, etc. in this publication does not imply, even in the absence of a specific statement, that such names are exempt from the
relevant protective laws and regulations and therefore free for general use.
The publisher, the authors and the editors are safe to assume that the advice and information in this book
are believed to be true and accurate at the date of publication. Neither the publisher nor the authors or the
editors give a warranty, expressed or implied, with respect to the material contained herein or for any
errors or omissions that may have been made. The publisher remains neutral with regard to jurisdictional
claims in published maps and institutional affiliations.

This Springer imprint is published by the registered company Springer Nature Switzerland AG
The registered company address is: Gewerbestrasse 11, 6330 Cham, Switzerland

If disposing of this product, please recycle the paper.

I attended primary and secondary school in Mexico City. I was already fascinated by science before entering high school; I still remember my excitement when I first looked at paramecia and amoebas through a rather primitive toy microscope.

José Mario Molina-Pasquel Henríquez

The microscopy under the gaze of an artist

Chinese ink painting

Fabiola García-Zamorategui

Preface by Ataúlfo Martínez-Torres

The first time you observe a sample under a microscope may have various profound and extreme effects in your subconscious. Visual contact with the world hidden to the naked eye is a powerful experience, often from the young elementary student to the research scientist alike.

Remember drawing your first mitotic cell? It is in these traces where creativity is often born. These moments have the potential to galvanize the creation of beautiful artworks, awesome narratives, and novel explanations to natural phenomena.

Light is a wave, a particle, a photon we detect in our retina with specialized photosensitive cells. Light is also fun and interesting. Light is exciting to experiment with. You can perceive the joy of Newton experimenting with prisms in his portrait by J.A. Houston. The invention of the microscope brought light to the domain of life science. We use the microscope to experience the microcosms beyond the limits of our physiological resolution, and through it, we also create art.

Our senses are inherently unreliable. That is why it is so valuable that we have access to instruments that allow us to enhance them and that serve as analytical tools in the precise description of the phenomena that we intend to define. Thus, the microscope is a fundamental tool for discovery.

As we harness new technologies, our sensory perceptions expand, revealing phenomena that would be impossible to grasp without specialized instruments. The tradition of using microscopes to create artistic illustrations has deep roots, exemplified by the exquisite engravings of Christopher Wren and Robert Hooke. Hooke's groundbreaking work, *Micrographia*, beautifully illustrates the intersection of science and art. Its meticulous descriptions and quantitative analyses of microscopic specimens, paired with detailed illustrations, have established it as a cornerstone in the evolution of scientific and artistic innovation, continuing to resonate today.

Following this rich tradition, this book presents a stunning collection of images captured through various techniques, showcasing the intricate details of microscopic specimens and the different types of microscopes utilized in scientific research. It also features portraits of renowned researchers who have leveraged the microscope to explore our microcosm and enhance our understanding of light and perception.

This work offers a fascinating exploration of both historical and modern microscopes, prompting reflections on the evolution of technological advancements aimed at pushing the boundaries of resolution. We now live in an era where super-resolution microscopy has surpassed the limits of light diffraction, enabling us to capture images previously deemed unattainable.

Each chapter is complemented by technical notes that provide valuable insights into the mechanisms driving these precise instruments. These notes serve as an invitation to delve deeper into the workings of the equipment, enhancing our understanding of how to optimize its use and maximize its applications.

One of the standout features of this collection is the diverse array of microscopy techniques employed to observe the specimens. Highlights include illustrations of brain tissues and cells stained with simple dyes, which reveal cellular components and organization, as well as advanced techniques utilizing fluorescent molecules and specially adapted microscopes for signal analysis. The advent of fluorescent proteins and other molecules has ushered in a renaissance in microscopy, allowing us to not only observe intricate details of cell morphology and organelles but also to study their functional dynamics in living systems.

Finally, the images captured with electron microscopes -both transmission electron microscopy (TEM) and scanning electron microscopy (SEM), challenge us to consider what lies beyond the resolution limits imposed by millions of years of evolution. Within these images, we encounter new aesthetic experiences, translating these perceptions into innovative artistic expressions.

Thus, this book represents a meticulous and enlightening approach to scientific knowledge, beautifully merging the wonders of the natural world in its smallest components—revealed to us through the most widely used instrument in research laboratories: the microscope.

Neurobiology Institute
National Autonomous University of Mexico
Queretaro, Juriquilla, Queretaro, Mexico

Ataúlfo Martínez-Torres

Preface

Our proposal, rooted in both literature and personal experience, seeks to convey the technical principles of six different microscopes in a clear and concise manner. Unlike traditional microscopy textbooks or specialized works, this book offers an engaging introduction to the world of conventional microscopy. We envisioned an audience that includes newcomers to the scientific field, particularly in microscopy, while also considering intermediate readers. Ambitiously, we believe it may even benefit specialists who, deeply immersed in their work, might be unaware of certain specific information presented here—details that are often difficult to find in conventional textbooks, especially regarding microscopes that fall outside their field of expertise.

This book harmoniously combines artistic creations—including Chinese ink drawings, watercolors, photographs, micrographs, and computer illustrations—with succinct academic technical content. Our aim is to make the reading experience enjoyable and to facilitate understanding of each section. We hope to strike a balance between the realms of art, science, and academic discourse. A book with these characteristics appears to have no precedent in the field of microscopy.

Contained in these pages, you will find eight chapters: one providing a brief history of microscopy, another outlining the foundational concepts necessary for understanding the subsequent sections, and the following six, focused on brighfield, widefield, single-photon confocal, two-photon confocal, scanning electron and transmission electron microscopes.

This work is the result of a collaborative effort among talented colleagues: Carmen Reyes Luna, language, content, and style editor, who ensured the coherence of the text through her translation of most sections from Spanish to English and provided grammatical corrections, as well as source validation of the many scientific facts found in these pages in a procedure similar to the double-blind system to ensure the objectivity of the information; Fabiola García-Zamorategui, an artist who beautifully illustrated the faces and microscopes featured in this book using Chinese ink and watercolor; Ricardo Piña-Muñoz, who co-created the technical illustrations with me, striving to provide visually appealing representations for each section; Alfonso Cárabez-Trejo, my mentor in electron microscopy, who graciously contributed his expertise and high-quality electron micrographs for the relevant sections; and Yazmín Ramiro-Cortés, a specialist in two-photon microscopy and mentor to Gerardo R. Perera-Murcia, who enthusiastically collaborated on the two-photon section by providing remarkable micrographs for that chapter. Lastly, we are grateful for the support and insightful guidance of Merry Stuber, the editor at Springer Nature.

Now that we have completed this work, we, the authors of the book you hold in your hands, sincerely hope you enjoy its content and find it to be the perfect companion for your favorite cup of coffee or tea, allowing you to savor both its aroma and flavor. We recommend that you read this book in sequence to better understand the chapters and to ensure that you do not find some of the terms or elements that make up the microscopes presented here complicated. In case this is not possible, we have also added a glossary of technical terms at the end of this book to support your reading.

On behalf of my colleagues

Mexico City, Mexico Abraham Rosas-Arellano

Acknowledgments

To you, the others behind the scenes, who with your help also collaborated to make this book possible.

- **Meng Daniel James Smith**, for support with proofreading
- **Isabel Acosta-Galeana** and **Martha Elizabeth Montané-Romero**, for early suggestions for the project
- **Lorelei Ayala-Guerrero**, for the support with the GFAP immunostaining sample preparation
- **Lorena Saragoni-Cisterna**, for the support with the Golgi stain sample preparation
- **Leticia Robles-Martínez**, for the support with the technical review of the final format of figures
- **Dulce Olivia López-Rivera**, for technical assistance in the acquisition of laser light micrograph, Fig. 5.4
- **Isaac Chávez-Blancas**, for format suggestions, figures of Chap. 2
- **Laura Marianna Cano-Mateo,** for technical support, Fig. 5.9
- **Ana Lucia Tovar-Álvarez,** for technical support on micrographs: Figs. 7.1 and 7.11
- **Veronica Garrido-Bazan,** for support in the Fig. 7.14
- ZEISS Mexico specially to:
  - **Rubén Hinojosa-Puga**
  - **Alejandro Olvera-Ordoñez**
  - **Diego Andrés González-Restrepo**
  - **Paula Jiménez-Miranda**
  - **Ángel Mauricio Ramírez-Gutiérrez**
  - **Hernán Gonzalo Gerez (Chair)**
- IFC-UNAM-Computing unit specially to:
  - **Ana María Escalante Gonzalbo**
  - **Gerardo Coello Coutiño**
  - **Ivette Rosas Arciniega**
  - **Juan Manuel Barbosa Castillo**

- IFC-UNAM-Library:
  - **Sandra Moncada Hernández**
  - **Javier Gallegos Infante**
- AR-A thank and recognizes to his microscopy teachers:
  - **Alfonso Cárabez-Trejo,** microscopy basis and transmission electron microscopy
  - **Lourdes Palma-Tirado,** transmission electron microscopy
  - **John Heuser,** transmission electron microscopy
  - **Ana Lucia Tovar-Álvarez,** scanning electron microscopy
  - **Nydia Hernández-Ríos,** confocal microscopy
  - **Alejandro Olvera-Ordoñez,** confocal microscopy
  - **Juan Riesgo-Escobar,** confocal microscopy
  - **Luis Vaca-Domínguez,** confocal microscopy

Introduction

The book you are holding in your hands is a unique scientific and cultural blend formed within an often contradictory set of circumstances at the onset of the second millennium (CE) in Mexico. This wonderful land inhabited by multiple native peoples received strong influences from the arrival of the Spanish in the sixteenth century. Its inhabitants fought and gained their independence 300 years later; armed struggles continued with mixed outcomes, first against other opportunistic invaders and then with a revolution against the dictatorial government of Porfirio Díaz. Systematic knowledge held by the natives was effectively destroyed in four centuries that saw limited appreciation for science.

How is it then that we have received a book summarizing in few pages the workings of the most valuable instrument available to biologists in its various forms? A book illustrated with exceptional original photographs of what can be made visible from the microcosm of which we are constructed (and surrounded) and with appealing caricatures to help the younger generations approach complex machinery without being discouraged? Pages with original reproductions in Chinese ink of portraits of the scientists and technologists that offered humanity instrumentation to look where our predecessors were unable to penetrate? How were these cultural gems born from a first-class artist after she met the authors?

During the past century, Mexico hosted waves of political refugees because of brutal conflicts in Spain and other parts of Latin America. These migrants were often well-educated, socially progressive individuals and, together with parallel inputs from the powers of the time (such as the United States of America and the Soviet Union) and local efforts, contributed a great deal to the establishment of scientific institutions in the country. Most emblematic in the Pantheon of Mexican scientific institutions are the National Autonomous University of Mexico and the Center for Research and Advanced Studies of the National Polytechnic Institute, where the book's main author, as thousands of his most illustrious contemporaries, received their education. Importantly, a universal system for broad provision of student and professional stipends is granted by the National Council of Technology and Science (recently upgraded into a Ministry of Science, Humanities, Technology and Innovation). These stipends, coupled to academic freedom and employment stability for university professors and researchers, have allowed for the development of a thriving educational and intellectual community, only constrained because of lags in technological infrastructure and development.

This book is a token of modern Mexico, where science and the arts can flourish together. It has been designed for young readers eager to learn how the microscopes they encounter in laboratories function, those curious about the key innovations that propelled microscopy forward with each new breakthrough, or anyone interested in exploring the diverse forms and shapes of cells through captivating images. The explanatory diagrams primarily target students but are also accessible to the general public. Furthermore, there's enough detail to engage specialists, who may uncover insights into the technical aspects and possibilities offered by various types of microscopes, potentially using this book as a resource in their own teaching to inspire the next generation of microscopists. Certainly, all readers will appreciate the book's micrographs and drawings.

A book that singularly unites so many attributes requires no further explanation to justify its existence. The expert microscopist behind its conception spent his formative years in Ricardo Miledi's laboratory. Miledi, who among other things has a claim to the discovery that calcium is required for neuronal connectivity, was a founding member of the Institute of Neurobiology in Juriquilla, Querétaro. In the same institute, the book's main author enjoyed close mentorship from Alfonso Cárabez-Trejo, who instilled in him an appreciation of electron microscopy, working side by side with Lourdes Palma-Tirado. Some of Alfonso's own images are included in this book.

My own arrival to Juriquilla, around 18 years ago, where I first met with the author, coincided with the acquisition of a state-of-the-art confocal microscope, which he used to map extrasynaptic neurotransmitter receptors in his doctoral thesis, becoming an expert. During a postdoctoral period at the Austral University of Chile, he documented the mispositioning of these receptors within neuronal synaptic connections in the aging or diseased brain. In Chile, the author met and collaborated with Ricardo Piña-Muñoz, a key figure behind the book's explanatory diagrams. In this same period of his flourishing career, he also introduced methodological improvements for standard sample preparation protocols (presently used by numerous researchers worldwide) that reduce optical noise (or "blur") attributable to chemical treatments inherent to conventional microscopy procedures.

With these enviable scientific achievements, the Institute of Cellular Physiology is particularly fortunate to count with the presence of the book's principal author in its Microscopy Unit, working tirelessly with students and obtaining high-quality images for the most diverse of projects. The reader must also wonder how the extra time arose to collect and compile the information we find in the following pages, from the history of microscopy, the principles of magnification using lenses, the inventors of various types of conventional microscopes, and the explanations of principles and technologies that support them. Due to wonderful coincidence, the author met the artist Fabiola García-Zamorategui. I recall his enthusiasm, which at the time I could not understand, for I hadn't had the opportunity to see the paintings, and the publisher encouraged them to embark onto this collaborative project. Yazmín Ramiro-Cortés and Gerardo R. Perera-Murcia are the experts on two-photon microscopy. Carmen Reyes Luna, who at the time worked with Dr. Ramiro, assumed the supervision and translation of the text. Their contributions are duly acknowledged throughout the book.

The perfectionist living inside a microscopist will become evident as readers focus their gaze on the selection of photographs offered herein. Still, an introduction should also mention that which might be missing from this ambitious effort to explain and summarize microscopy.

I often drink my morning coffee in a Stanford synchrotron mug, a gift I received from the author when he performed X-ray fluorescence imaging in that facility. This was my laboratory's first introduction to the majestic world of synchrotrons, which includes a considerable set of sophisticated microscopical techniques that permit visualization, pixel per pixel of the elemental content of a sample. The book will not cover such recent advances of instrumentation, its declared purpose is to teach and showcase the microscopes commonly available in universities and institutes world-wide. Instruments I would have liked to see mentioned include the LightSheet microscope (for its improved ability to monitor the development of living embryos) the cryo-electron microscope, and perhaps the super-resolution microscope (which could have claimed its own chapter following a Nobel Prize). In the grand scheme of the book, these omissions are rather minor.

Some comments are offered on the fascinating, and continuously debated, nature of light, including how laser beams are formed. However, to receive a treatise into the nature of light—the main form of how scientific information enters our con-science (just think of what you have learned in your lifetime through interaction with light, including while reading these lines)—you may need to look elsewhere. Nevertheless, it is my hope that an original attempt from the author to make an image of a laser beam, included in these pages, could kick-start a new research project in years to come.

From placing microscope lenses on a chess board to a fluorescent planting field, Van Gogh, Siamese and Big Bang cells, from furious fire and green splash to the climbing arm of a house cricket or a brain-storming session of neurons, prepare yourself for a visual feast into the microcosm. How was it that Maria Göppert-Mayer's Ph.D. thesis on photo-physics in the 1930s became key 60 years later to develop a new type of microscope? What did Maria look like? scanning electron microscopy, transmission electron microscopy, and confocal and two-photon microscopy will all become familiar once you have read the book.

While looking at the beautiful illustrations, do not forget to consider how it is that we can see these microscopic wonders, explained using simple words and friendly diagrams. Hopefully, next time you use the microscope you will make wiser decisions on its handling. Advice I received early on was to consider that a good image that clearly communicates a message lasts a very long time and it is therefore fair to dedicate considerable care to make it as good as possible. But also, always remember that what seems apparent may not always be true.

Thank you to all the authors, the painter, and the publisher for this work, for creating the book that is now here for you to enjoy.

Cinvestav   Fanis Missirlis
Mexico City, Mexico

Contents

1	**The Origins**	1
	References	10
2	**The Basis**	13
	2.1 The Nature of Light	14
	2.2 The Duality of Light: Wave and Particle	15
	2.2.1 Refraction of Light	19
	2.2.2 Reflection of Light	20
	2.2.3 Lens Geometry	22
	2.2.4 Lens–Light Interaction (Convergence, Divergence, and Dispersion of Light)	23
	2.2.5 Lens Aberrations	25
	References	32
3	**The Brightfield Microscope**	33
	3.1 When Should a Brightfield Microscope Be Used?	47
	References	53
4	**The Widefield Microscope**	55
	4.1 When Should a Widefield Microscope Be Used?	69
	References	70
5	**The Confocal Microscope**	71
	5.1 When Should a Single-Photon Confocal Microscope Be Used?	95
	References	96
6	**The Two-Photon Microscope**	99
	6.1 The Two-Photon Microscope	100
	6.2 Infrared Light	104
	6.3 Deep Imaging	105
	6.4 In Vivo Imaging	106
	6.5 When Should a Two-Photon Microscope Be Used?	111
	References	112

7 The Scanning Electron Microscope 115

    7.1   When Should a Scanning Electron Microscope
          (SEM) Be Used? 131

    References 135

8 The Transmission Electron Microscope 137

    8.1   When Should a Transmission Electron Microscope (TEM)
          Be Used? 153

    References 154

Glossary 155

About the Authors

Abraham Rosas-Arellano, Ph.D. received his Ph.D. in Biomedical Sciences from Neurobiology Institute of the National Autonomous University of Mexico (UNAM). He has two bachelor degrees, one of them in Biology and the other one in OB-Gyn Nursing, as well as a master's degree in Science, all of which were completed at UNAM. During his career he received an extense and comprehensive training in techniques and methodologies focused on fluorescent and electron microscopy. He has been awarded various scholarships from CONAHCYT, AMUHJ, and from the Society for Neuroscience. He is a member of the National System of Researchers in Mexico. He completed his postdoctoral research at the Marine Biological Laboratory in Woods Hole Massachusetts (USA), Cinvestav-IPN in Zacatenco (Mexico City, Mexico), and at the Universidad Austral de Chile in Isla Teja (Chile). Additionally, he's carried research stays at the synchrotrons of Stanford (SLAC National Accelerator Laboratory) and in Paris (Synchrotron Soleil, Paris-Saclay). Currently, he is part of the Imaging Unit staff at the Institute of Cellular Physiology at UNAM as specialist in brightfield, widefield, confocal, transmission electron, and scanning electron microscopy. He has been invited as a guest professor to teach specialized theoretical and practical microscopy courses both in and outside of Mexico.

Carmen Reyes Luna, MSc was born and raised in Mexico City. She obtained a Bachelor of Science degree in Pharmacology and a Master of Science degree with Distinction in Precision Medicine at the University of Manchester (United Kingdom), same institution where she obtained "The Stellify Award" for volunteering, leadership, and ethical roles. She participated in her first research project with Dr. Ayse Latif, investigating monocarboxylate transporters in endometrial cancer. After her graduate studies she returned to Mexico City with the intention of developing her research skills in neuroscience. She became part of the laboratory of Dr. Ramiro-Cortes at Institute of Cellular Physiology of the National Autonomous University of Mexico where she got involved in the study of network and engram aberrations present in animal models of Phelan-McDermid Syndrome. Currently, she works as a scientist for Metsera in London, dealing with experimental treatments for diabetes and obesity. Her personal research interests, galvanized by being awarded a Psychedelic Science 2023 scholarship, include mental health and disease, neuroplasticity, and neurodegeneration.

Fabiola García-Zamorategui is a Mexican Pharmacobiological Chemist and Artist dedicated to body tattooing since 2014, specializing in "Black & Grey Realism," sacred geometry with dotwork, and "Blackwork." She has built her career in studios such as "Studio Honour & Pride," "Mala Sangre Tattoo Crew," "Red Cloud Mavericks Tattoo Studio," "Alacrania," "Arcadia," "Red," "Cloud Mavericks," "Diabolink," and currently at "3 Seis Tattoo & Art Collective," where she is known as Yinfayang Tattoo or Fabiola Zamorategui. She creates paintings using the "Chinese ink wash" technique and sculptures using papier-mâché, moldable dough, and clay techniques. Recently, she has begun working as a book illustrator for various genres.

Ricardo Piña-Muñoz, Ph.D. received his Ph.D. in Sciences with a specialization in Molecular Biology, Cellular Biology, and Neurosciences from the University of Chile. He completed a postdoctoral fellowship in Neurosciences at the University of Santiago de Chile and a postdoctoral internship at the University of Erlangen-Nuremberg (Germany). He is currently a titular Professor in the Department of Biology at the Metropolitan University of Educational Sciences located in the commune of Ñuñoa, Chile, where he engages in teaching and research activities. His research focuses on topics related to cellular biology and neurophysiology in the Central and Peripheral Nervous Systems at the cellular and molecular levels.

Yazmín Ramiro-Cortés, Ph.D. is a researcher at the Institute of Cellular Physiology of the National Autonomous University of Mexico (UNAM) and is currently a member of the National System of Researchers. She studied Biology at the Meritorious Autonomous University of Puebla (BUAP) and then moved to Mexico City to pursue a Ph.D. in Biomedical Sciences at the Institute of Cellular Physiology, UNAM. She subsequently completed a postdoctoral stay at the Gulbenkian Institute of Science at the Champalimaud Foundation in Portugal in the laboratory of Dr. Inbal Israely. In 2014 she joined the Institute of Cellular Physiology at UNAM, where she is a researcher and is also in charge of providing the service for the use and management of the two-photon excitation microscope. In her laboratory, the neural bases of autism spectrum disorder are studied in murine models, through the analysis of neuronal activity in vivo and the study of morphological changes of individual synapses and dendritic spines, as well as synaptic plasticity and neuronal activity related to learning.

Gerardo Rodrigo Perera-Murcia, MSc is a biologist who graduated from the Faculty of Sciences at the National Autonomous University of Mexico (UNAM). He also completed his master's degree and is currently finishing his doctorate, both master's and Ph.D. degrees under the supervision of Dr. Ramiro-Cortes at Institute of Cellular Physiology of the UNAM. His research is focused on analyzing changes in synaptic integration processes in models of autism at the Institute of Cellular Physiology with Dr. Yazmín Ramiro, who works on structural synaptic plasticity and neuronal activity in the visual cortex. Currently, he is an academic technician at the laboratory of Dr. Julio Moran, who studies neuronal death mechanisms at the Institute of Cellular Physiology, UNAM. In his spare time, he is a neuroscience lecturer at the Monterrey Institute of Technology and Higher Education, also known as Technological Institute of Monterrey (campus Mexico City). Currently, his research interests include synaptic plasticity, two-photon microscopy, hippocampal physiology, structural plasticity, and neuronal cell death.

Alfonso Cárabez-Trejo, M.D., Ph.D. a medical surgeon from the School of Medicine at National Autonomous University of Mexico (UNAM). He holds a Ph.D. with a specialization in Biochemistry from the Faculty of Chemistry at UNAM. His various academic positions at the School of Medicine at UNAM include Laboratory Instructor in Biochemistry, Teaching Assistant, Adjunct Professor, Associate Professor, Course Lecturer, and main/titular Professor. Additionally, he served as a Principal Investigator at UNAM in the Department of Experimental Biology at the Institute of Biology, Institute of Cellular Physiology (IFC), and later at the Department of Developmental Neurobiology and Neurophysiology at the Institute of Neurobiology (INB), Campus Juriquilla. Internationally, he was Associate Researcher at the Johns Hopkins School of Medicine, Research Assistant at Cornell University's Division of Biological Sciences, and Visiting Professor at the University of Tennessee. His research focused on bioenergetics, particularly mitochondrial ATPase, and micromorphology. For this, he relied on transmission electron microscopy and scanning electron microscopy. He was instrumental in excellency trainings of electron microscopy and in the establishment of microscopy units in Mexico, including the Electron Microscopy Unit at Cinvestav Irapuato, Microscopy Unit at INB—UNAM, and the Imaging Unit at IFC—UNAM. He is now retired.

The Origins

1

People who look for the first time through a microscope say, "Now I see this, and then I see that," and even a skilled observer can be fooled. On these observations I have spent more time than many will believe, [making microscopic observations] but I have done them with joy, and I have taken no notice of those who have said, "Why take so much trouble," and "What good is it?"

Anton van Leeuwenhoek

The resolution of a microscope is the power that it has to magnify, with exquisite detail and visual quality, specimens that are otherwise imperceptible to the naked human eye. Resolution is dependent on three main aspects: (1) the energetic properties of the light projected onto the sample, (2) the combination of the characteristics of the carving of the lenses, and (3) the precision in the coordination between them.

The first ever historical record of glass being used as a magnifying lens comes from the time of the Assyrian empire. Known as the lens of Nimrud or as the rock crystal of Nineveh (Fig. 1.1), it is presumed that this rock crystal could have had other uses distinct from object magnification. Other records indicate that the use of the water flask to enlarge objects may have been known in ancient Egypt and was probably known to jewelers in the region.

The original application of superimposed lenses for object magnification can be attributed to Zacharias Janssen (Fig. 1.2). His invention consisted of an instrument with two lenses; at the fore, the first lens was positioned in front of the object to be observed, while at the rear, the second lens was closer to the observer. Both lenses were interconnected by a series of metallic cylinders (Fig. 1.3). To properly focus an object, the observer could slide the metallic cylinders, achieving dynamic focusing suitable for the observation of many different types of samples. Based on its double conjugate lenses, this device has been considered the first compound microscope.

© The Author(s), under exclusive license to Springer Nature Switzerland AG 2025
A. Rosas-Arellano et al., *Microscopic Wonders*,
https://doi.org/10.1007/978-3-031-92559-7_1

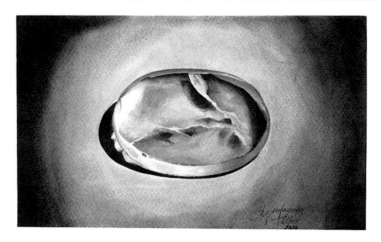

Fig. 1.1 Nimrud lens
With undeniable optical characteristics and apparent use for object magnification, it has been recognized by diverse historical sources as the most ancient lens ever recovered by humanity. It was discovered in 1850 by Austen Henry Layard (1817–1894) in the Nineveh ruins, which was an ancient Assyrian city located in modern-day northern Iraq. The Nimrud lens was unveiled to the scientific community in 1853 by David Brewster (1781–1868) through the British Association for the Advancement of Science. The origin of this lens dates back between the years 750 and 710 BC; crafted from natural rock crystal, it possesses a rustic carving with one convex and one plane surface. With a focal length of 12 cm, the lens has a diameter of 1.25–1.63 cm and a thickness of 4.10–6.20 mm; it has been attributed a magnifying power of 3×. Despite its inherent optical characteristics, both physical and functional, there is no conclusive evidence regarding the reason for its creation or its use as a magnifying lens. The qualities of the oval piece as an object magnifier could be merely accidental, and several scholars have suggested different purposes for this item, including for jewelry, to generate fire by focusing sunlight beams, or as a component of an ancient telescope, which would make it the first telescopic piece ever found. Unfortunately, there are no other pieces of evidence that allow for this relationship to be established

However, there are some controversies regarding the paternity of this first compound microscope. One of them poses the possibility of Johan Janssen, Zacharias Janssen's father, being the actual inventor of the Janssen's microscope. This is sustained by the age of Zacharias, who was still a child in the year the instrument was dated back to. Also supporting this version of events is the fact that, when Zacharias was very young, Johan had a traveling show called "image magnifier." Additionally, there is a version of the story that attributes the invention of this microscope to Zacharias's grandfather, Hans Martens.

The controversy regarding the design of the first compound microscope extends beyond the Janssen family. It is also believed that Hans Lipperhey, a Wesel-born German spectacle maker, was the actual designer of the double lens conjugate. Lipperhey lived in the Netherlands and has been frequently associated with the Janssen family. However, this version is not exempt from controversy either, as it is believed that Lipperhey could have stolen the idea of superimposing two lenses by watching Zacharias Janssen play with a pair of them as a child. Regardless, research carried out around the year 1820 points to the Janssen family, most likely to

1 The Origins

Fig. 1.2 Zacharias Janssen
Zacharias was born in The Hague, the Netherlands, between 1580 and 1588 and died between 1632 and 1638. However, due to the loss of records during the Second World War bombings, the exact dates are unknown. He was known as a spectacle-maker, a profession that he learned from his parents—apparently more so from his mother, Maeyken Meertens, than from his father, Hans Janssen, who died when Zacharias was just a child

Zacharias himself, as the legitimate inventors. Nevertheless, Lipperhey also entered history as the father of the first telescope.

Regarding the naming of this invention, the first person to originally use the word microscope was Johann Giovanni Faber (Fig. 1.4), who employed it to refer to an instrument created by the Italian Galileo Galilei. This instrument consisted of a conjugate of three lenses and was originally referred to as occhiolino (which means little eye). Faber named the instrument based on two Greek words: "mikros," which means "small," and "skopein," which means "look" or "observe." Faber coined the term microscope to highlight the difference between this instrument and a telescope, employed to look in detail at what is large and distant. Since Galileo kept both a microscope and telescope in his study room, it was almost unavoidable that Faber saw them and ended up coining distinct words to distinguish them. After its naming, the microscope remained relevant for years before it was modernized, yet again.

Considered an icon of microscopy, in addition to one of the most representative figures in science, Anton van Leeuwenhoek (Fig. 1.5) was a textile and cloth merchant at some point in his life. Due to his work, he had constant interaction with lenses, using them to analyze the threads of fabrics. Given the tiny size of the lenses he employed, he also learned to design and craft his own devices from brass, copper, silver, and gold to hold the small lenses in place. The highest-resolution lens preserved from Leeuwenhoek measures less than 2 mm in diameter, offers a magnification of 266×, and has a resolution of 1.35 microns. Its focal length is 0.9 mm,

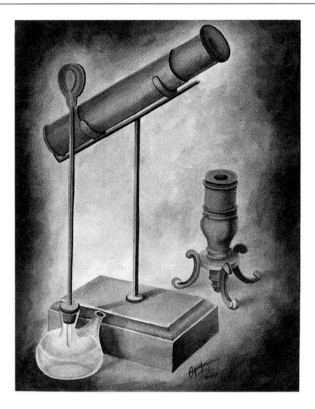

Fig. 1.3 Janssen's microscope
This microscope has been dated back to 1595. It consisted of three brass tubes and two lenses. Two of the drawtubes have slightly different diameters—the largest one not surpassing 6 cm—so they can slide into each other while being held in place by an outer third tube. When closed, it is 25 cm long, but when fully extended, it reaches 45 cm. Inserted in each extreme of the flanking tubes, one lens functions as a biconvex eyepiece lens (the frontal lens), and the other lens functions as a planoconvex objective lens (the posterior lens). As a consequence of the drawtube's displacement, the distance variations between the lenses allowed for multiple focal points, achieving magnifications between 3× and 9×. Due to its historical context, the poor quality of glass, and its rudimentary carving, this microscope produces low-resolution images with significant optical aberrations. Despite these flaws, the Janssen's microscope posed a major scientific and technological advance due to the use of conjugate lenses, as opposed to the contemporary employment of a single lens for magnification purposes

meaning the sample had to be positioned precisely at that distance. Mounting such a tiny lens onto a metal frame without damaging it must have been, without a doubt, a challenging task. Leeuwenhoek's innovations in lens design improved specimen visibility, increased magnification, enhanced resolution, and broadened the utility of his microscopes for observing both liquid and dry samples. As a microscopist, he constructed over 500 microscopes (at least 350 of which were completed), specifically designed to observe three-dimensional objects that could be moved or rotated along an axis, thanks to the accessories he created.

Fig. 1.4 Giovanni Faber
German physician, anatomist, and botanist; he met Galileo Galilei at the Accademia dei Lincei in Rome, which was one of the most prestigious European scientific institutions. In 1625, for the first time, Faber used the word "microscope" to differentiate the employment of this instrument from its counterpart, the telescope. Faber coined the term a year after Galileo unveiled his microscope in the Lincei. The use of the word microscope led to the word "microscopy" to define the area of knowledge, which was subsequently followed by the term "microscopic" applied to the samples observed through these instruments. Microscopes are currently considered essential in scientific research institutions, as it is almost impossible to fathom a laboratory that lacks them. They have even come to be known as one of the most representative symbols of scientific research

Additionally, these microscopes maintained a constant but adjustable distance between the lens and the sample, allowing the focus to be fine-tuned and held steady for prolonged observation of intricate details. Leeuwenhoek performed countless microdissections on animals and plants, documenting both their external and internal structures. He observed and described a range of subjects, including grain meals; fruit seeds; coconuts; reproductive organs and cells; muscles; scales; the nervous system; hair; skin; and tendons, all at magnifications that were unprecedented at the time. He also examined and described inorganic samples, such as minerals and salts.

For his most famous microscope, Leeuwenhoek created an innovative design of a microscope that was just a few centimeters in size (Fig. 1.6). This microscope consisted of a lens with exceptional carving, fashioned by Leeuwenhoek himself, and two screws that could be turned to adjust the position of the sample, consequently providing dynamic adjustment of the focal point. Additionally, he suggested

Fig. 1.5 Anton van Leeuwenhoek
Native of Delft, the Netherlands, Leeuwenhoek (1632–1723) lacked higher education, yet managed to surprise the world with his curiosity and perseverance. The lack of an institutional affiliation did not limit his scientific contributions, which led to him being considered one of the pillars of microscopy and micromorphology, as well as a key precursor to cellular biology and microbiology. To date, Leeuwenhoek is considered a key precursor in more than ten disciplines. His achievements garnered him invitations to join the Royal Society of London and the Academy of Sciences of Paris. In addition to his invaluable findings, Leeuwenhoek stood out due to the outstanding quality of carving in his microscope lenses, which he fashioned and manufactured himself. These lenses were extremely hard to replicate, and for many years, his carving procedures, methods, and techniques to observe minute samples were guarded as precious secrets. The magnifying power achieved by Leeuwenhoeks' microscopes was close to 300 times the size of the original sample (300×), something never seen before his time. He built hundreds of microscopes that ranged from single-lens ones to some containing up to three lenses. He was, without a doubt, one of the first scientists to ever observe single-celled organisms, which he named "animalcules." It is highly likely that Leeuwenhoek was the first person to ever record the existence of erythrocytes, spermatozoa, and bacteria, among many other inorganic entities

many techniques to improve microscopic observation, such as dark field microscopy and the use of *mica sheets* for fragile samples (which had the same function as current slides), among others.

Like Leeuwenhoek, another pioneer in the field of microscopy was Robert Hooke, a man who, interestingly, at one time called Leeuwenhoek the microscope's "single votary" (Fig. 1.7). Due to the poor recognition of his numerous contributions, Hooke has been considered one of the biggest and most brilliant forgotten geniuses in scientific history. Moreover, his contributions were intentionally

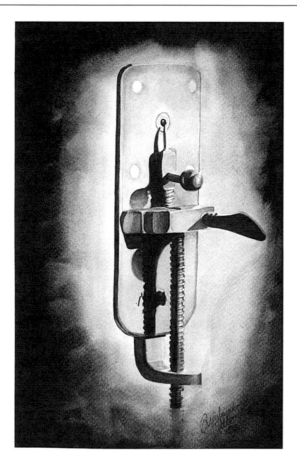

Fig. 1.6 Leeuwenhoek's microscope
This is a highly original and beautiful microscope that stands out for its innovative design. It consisted of a small bronze plate with an orifice in which a biconvex lens could be perfectly attached. In front of this lens, a specimen could be placed on a pin near a structure that rotated the sample 360° along its vertical axis. This was supported by a screw that could be turned to adjust the position of the sample in its inferior and superior axis. At its extreme, there was a flat head, where it could be held by hand. To focus the image, the viewer had to hold the microscope and place it close to their eye, adjusting the focal length millimetrically using a second screw that moved the specimen in the Z axis. Both adjustments were made possible using a double-threaded hole bracket where the screws moved when rotated

obscured by Issac Newton, with whom he maintained a fierce rivalry. In reality, Hooke made valuable contributions in various scientific fields. For instance, he formulated the theory of planetary motion; discovered a star in the Orion constellation; optimized the telegraph; adapted springs and other components for watches; developed the Gregorian telescope; described a binary star on the planet Uranus; invented the barometer, hygrometer, and anemometer (which measures wind speed and pressure); established the freezing point of water; described the elastic properties of materials when deformed by a force (Hooke's Law); discovered the existence of

Fig. 1.7 Robert Hooke
Hooke (1635–1703) was born in Freshwater, England. He attended the Christ Church College at Oxford but did not earn any degrees. However, due to his scientific contributions, he was appointed as a member of the Royal Society of London, where he held roles, first as a curator of experiments and finally as a fellow. This Chinese ink painting is a replica of the reconstruction work by Rita Greer, finalized in 2003. This reconstruction could not have been easy to accomplish, as records indicate that the only painting that ever existed depicting the physical aspect of Hooke was either burned or disposed of by Isaac Newton, who did so by taking advantage of his role as the president of the Royal Society of London. Newton repeatedly attempted to overshadow Hookes' scientific contributions after the death of the researcher. The fierce rivalry between Hooke and Newton was mainly due to a strong dispute regarding intellectual property over the Law of Universal Gravitation, which, as we know, ended up being attributed solely to Newton. A standout among other disputes was Hooke's proposal regarding the wave-like behavior of light, which opposed Newton's particle theory of light. Moreover, Newton tended to have a withdrawn personality, and in the Royal Society, Hooke had a nullifying and overly critical attitude toward him

extinct species; and designed hospitals, buildings, churches, and the Royal Greenwich Observatory. Moreover, Hooke also described the diffraction of light as a physical phenomenon and proposed that light had an inherent wave-like behavior.

Regardless of the attempts of Newton to overshadow his contributions, the name of Robert Hooke holds special importance in the history of science, as he wrote one of the best texts in the field of microscopy: *Micrographia*. This text is an impressive treatise on microscopy. Thanks to the observation and description of a magnified piece of cork, Hooke was able to employ the term cell in his book in a pioneering fashion. The porous texture of the cork under the microscope appeared to be empty polyhedral cavities, which he associated with cells, leading him to coin the term "cellulae." Since then, the word cell has been used to refer to the fundamental unit of all biological tissues.

1 The Origins

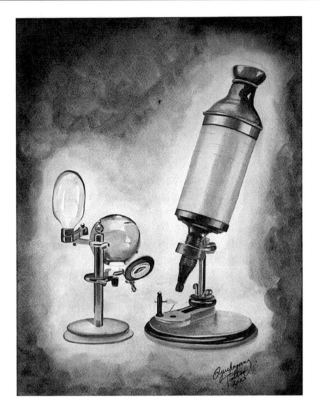

Fig. 1.8 Hooke's microscope
The schematic design of Hooke's microscope is the centerpiece of his iconic illustrated microscopy book. With refined leather engravings, this microscope is regarded as one of the most beautiful and elegant in the history of microscopy. This delicate instrument consists of four drawtubes contained within a 15 cm long cylinder; a plano-convex lens inlay can be found in one extreme, while in the other extreme, a biconvex objective lens rests. The superior end takes the shape of a cup, which serves as an eye rest and, importantly, maintains an optimal focal distance between the viewer's eye and the eyepiece lens. This microscope magnified between 20 and 50 times. Like modern microscopes, this instrument has a very short focal length. In 1665, employing this microscope, Hooke finished his first illustrated microscopy book: *Micrographia*. Published by the Royal Society of London, it contains dozens of photomicrographs hand-drawn by Hooke. This book has been rightfully recognized as a masterpiece of illustrated microscopy

Micrographia generated fascination as soon as it was published. This illustrated book includes micrographs of inorganic samples, fossils, plants, and animals—all hand-drawn by Hooke himself. The book positioned Hooke as an irreplaceable figure in the history of microscopy. Undoubtedly, the document must have emerged under the fascinated gaze of the scientist seeing specimens in detail, who employed in his research a microscope for which he planned the optics and design, alongside the Londoner Christopher Cock.

Hooke's microscope (Fig. 1.8) has an elegant and innovative finish. It possesses characteristics distinctive of all modern microscopes: a source of illumination, dynamic condenser, stage, sample holder, objective lens, and an eyepiece tube ending with a monocular lens system; all these elements are interconnected by means of a stand and a supporting mechanical arm that helps coordinate the focal adjustment and facilitate the operation of the system as a whole.

Over the years, microscopes have stood out for their varied and outstanding designs that range from rustic manufacturing; multiple sample holders; and detailed finishes in wood, copper, brass, and all kinds of metals or alloys. Microscope designs are always fashioned to favor the optimization of two central axes: the optic source and the light source.

The first designs stand out for their hand-carved lenses, such as the outstanding lens work carried out by Leeuwenhoek, whereas now impressive finishes are made with laser carving that can be accompanied by finishes that are resistant to high-energy light and are produced through a process known as photolithography.

Over the years, complementary materials with universal measures for microscopes have appeared. Some examples are slides, coverslips, and selective immersion media. Relevantly, many of the later models that followed Hooke's microscope already had components that could be easily and neatly grouped into three general categories: illumination, mechanical, and optical.

Currently, microscopy has strongly influenced scientific research, and there is no doubt that it is an indispensable field within it.

Further Reading

For historical generalities: Adeel (2016), Allen (2015), Bardell (2004), Gerhard (2021), Rochow and Tucker (1994), Sines and Sakellarakis (1987), Sushmasusik and Hayath (2015), Wells et al. (1931), and Wollman et al. (2015).

For Hooke: Ash (1998), Davidson (2009), Hooke (1961), Kemp (2003), Koyré (1952), Moxham (2016), Oldroyd (2003), Peters (2024), and Rowbury (2012).

For Leeuwenhoek: Anderson (2014), Karamanou et al. (2010), Kutschera (2023), Robertson (2023), and van Zuylen (1981).

For Nineveh lens: Barker (1930).

For Newton: Keynes (1980) and Koyré (1952).

References

Adeel AA (2016) A relic of the wellcome tropical research Laboratories in Khartoum (1903-34). Sudan J Paediatr 16(1):67–75

Allen T (2015) Microscopy: a very short introduction, very short introductions. Oxford Academic, Oxford

Anderson D (2014) Still going strong: Leeuwenhoek at eighty. Antonie Van Leeuwenhoek 106(1):3–26. https://doi.org/10.1007/s10482-014-0152-1

Ash C (1998) Hooke's microscope. Trends Microbiol 6(10):391. https://doi.org/10.1016/s0966-842x(98)01380-8

References

Bardell D (2004) The invention of the microscope. Bios 75(2):78–84

Barker WB (1930) Lens work of the ancients II: the Nineveh lens. Br J Physiol Opt 4:4–6

Davidson MW (2009) Pioneers in optics: Zacharias Janssen and Johannes Kepler. Microscopy Today 17(6):44–47. https://doi.org/10.1017/S1551929509991052

Gerhard C (2021) On the history, presence, and future of optics manufacturing. Micromachines (Basel) 12(6):675. https://doi.org/10.3390/mi12060675

Hooke R (1961) Micrographia (Royal Society of England, 1665, reprinted by Dover: New York)

Karamanou M, Poulakou-Rebelakou E, Tzetis M, Androutsos G (2010) Anton van Leeuwenhoek (1632-1723): father of micromorphology and discoverer of spermatozoa. Rev Argent Microbiol 42(4):311–314. https://doi.org/10.1590/S0325-75412010000400013

Kemp M (2003) Updating Hooke. Nature 424:255. https://doi.org/10.1038/424255b

Keynes M (1980) Sir Isaac Newton and his madness of 1692-93. Lancet 1(8167):529–530. https://doi.org/10.1016/s0140-6736(80)92777-4

Koyré A (1952) An unpublished letter of Robert Hooke to Isaac Newton. Isis 43(4):312–337

Kutschera U (2023) Antonie van Leeuwenhoek (1632-1723): master of fleas and father of microbiology. Microorganisms 11(8):1994. https://doi.org/10.3390/microorganisms11081994

Moxham N (2016) An experimental 'life' for an experimental life: Richard Waller's biography of Robert Hooke (1705). Br J Hist Sci 49(1):27–51

Oldroyd D (2003) Hooke, life and thinker. Nature 423:384–385. https://doi.org/10.1038/423384b

Peters WS (2024) Will the real Robert Hooke please stand up? Plant Cell 36:4680

Robertson LA (2023) Antoni Van Leeuwenhoek 1723–2023: a review to commemorate van Leeuwenhoek's death, 300 years ago. Antonie Van Leeuwenhoek 116:919–935

Rochow TG, Tucker PA (1994) A brief history of microscopy. In: Introduction to microscopy by means of light, electrons, X rays, or acoustics. Springer, Boston. https://doi.org/10.1007/978-1-4899-1513-9_1

Rowbury R (2012) Robert Hooke, 1635-1703. Sci Prog 95(Pt 3):238–254. https://doi.org/10.3184/003685012X13454653990042

Sines G, Sakellarakis YA (1987) Lenses in antiquity. Am J Archaeol 91(2):191–196. https://doi.org/10.2307/505216

Sushmasusik MS, Hayath S (2015) History of microscopes. Indian J Medn Alli Sci 3(3):170–179. https://doi.org/10.5958/2347-6206.2015.00034.5

van Zuylen J (1981) The microscopes of Antoni van Leeuwenhoek. J Microsc 121(Pt 3):309–328. https://doi.org/10.1111/j.1365-2818.1981.tb01227.x

Wells HG, Huxley J, Wells GP (1931) The science of life. The Waverley Book Company, ltd

Wollman AJ, Nudd R, Hedlund EG, Leake MC (2015) From Animaculum to single molecules: 300 years of the light microscope. Open Biol 5(4):150019. https://doi.org/10.1098/rsob.150019

The Basis

2

Have not the small particles of bodies certain powers, virtues or forces, by which they act at a distance, not only upon the rays of light for reflecting, refracting and reflecting them, but also upon one another for producing a great part of the phenomena of nature?

Isaac Newton

© The Author(s), under exclusive license to Springer Nature Switzerland AG 2025
A. Rosas-Arellano et al., *Microscopic Wonders*,
https://doi.org/10.1007/978-3-031-92559-7_2

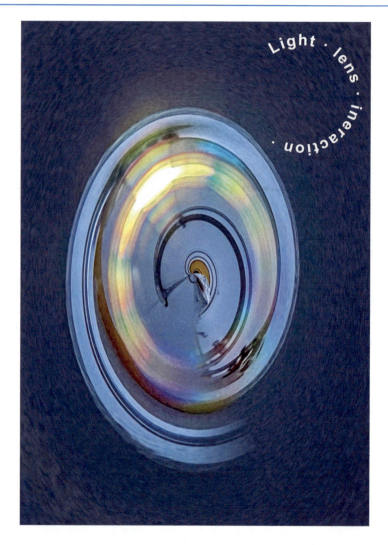

2.1 The Nature of Light

Curiosity regarding light's nature is as ancient as it is controversial. It took scientists hundreds of years to understand the physical characteristics that constitute light, and despite the passage of time and new scientific advances, there is still controversy about how much we truly know about it. Historically, the study of light was concentrated in two aspects of knowledge: (1) the one that conceived light as a compound of particles, known as the particle theory of light, and (2) the one that envisions light having a wave-like nature, named the wave theory of light. This dichotomy endured throughout time and was debated among some of the best minds in the history of humankind.

Early records speculating over the nature of light date back to the time of ancient Greece, where inductive reasoning prevailed over deductive reasoning. Aristotle (384–322 BC), for instance, postulated that light was not composed of matter; Democritus (460–370 BC) proposed that it was composed of small invisible particles; and Ptolemy (367–283 BC) established that light was a form of expression of energy.

In the times when modern science relied on hypothetical deductive thinking, a heated rivalry arose between the postulants of both theories. Among the scientists who described light as a particle were René Descartes (1596–1650 AC) and Isaac Newton (1643–1727 AC) (Fig. 2.1). The latter was the one who largely developed the corpuscular theory of light, which theorized light was composed of small corpuscles. This theory prevailed for a long time with very little willingness to refute it within the scientific community, which was probably related to Newton's privileged political position and outstanding scientific reputation within the Royal Society of London.

On the other hand, the scientific current that supported light's wave-like nature was mainly represented by Christiaan Huygens (1629–1695 AC), who is recognized as the creator of the wave theory of light; Robert Hooke (1635–1703 AC); Augustin-Jean Fresnel (1788–1827 AC); and of course, Thomas Young (1773–1829 AC, Fig. 2.2). The latter described the wave-like behavior using the "double-slit experiment"(Fig. 2.3). In 2002, Physics World (IOPscience) deemed this assay the most beautiful experiment in physics.

2.2 The Duality of Light: Wave and Particle

At the beginning of the twentieth century, Albert Einstein made incredibly new proposals regarding the nature of light at the quantum level. He pointed out that light, in addition to behaving like a wave, is also composed of quanta of energy (currently known as photons). This wave–particle duality was key in the discovery of the photoelectric effect that earned him the Nobel Prize in Physics in 1921.

Therefore, is light a wave or a particle? Under different conditions, light has been shown to behave like either of them. Consequently, despite Einstein's experimental descriptions, the scientific debate concerning the nature of light prevailed for many years due to substantial evidence of their duality appearing at different moments during recordings. It was not until 2015, when Fabrizio Carbone and his work group used a transmission electron microscope to photograph laser light, that it was demonstrated that light can behave simultaneously as a wave and particle (Fig. 2.4). Regardless of this experiment, speculation has not died. Does light naturally exist in this state of duality? Or are we, perhaps, unknowingly forcing it to behave like this in order to record and observe it, when its true nature is another one? What is light when it is not under our observation? Nonetheless, light's wave–particle nature produces interesting and varied consequences when exposed to different objects. This triad, wave–particle–object, is fundamental to understanding the bases of microscopy.

Fig. 2.1 Isaac Newton
Newton was born on December 25, 1642, in Woolsthorpe, England. Since childhood, he proved interested and skilled when working with mechanical models. Considered one of the geniuses responsible for key watershed moments in science, his popularity is often compared to that of scientists of the stature of Marie Curie and Albert Einstein. He is recognized as the man who developed one of the most remarkable methods for the study of movement, acceleration, and speed: Calculus, which became one of the most employed procedures in several fields, including physics, mathematics, technology, economics, administration, and biological sciences. He authored basic laws of mechanics and universal gravitation, as well as describing the nature of light in his book: "Opticks: or, a treatise of the Reflexions, Refractions, Inflexions and Colors of Light." His outstanding scientific career led to his appointment as president of the Royal Society of London in 1703, a position he spent 24 years being annually re-elected for. He died in March 1727. Postmortem studies detected traces of gold, arsenic, antimony, mercury, and lead in his hair, suggesting metal poisoning from his alchemy experiments. It is believed that the toxic effects of the metals are to blame for the rapid decay of his mental health, reflected by his difficult personality, which unfortunately remains a characteristic trait that the scientist is remembered by

2.2 The Duality of Light: Wave and Particle

Fig. 2.2 Thomas Young
Born in 1773 in Milverton, Somersetshire, Great Britain and diseased in 1829 in London. Young is considered the epitome of modern interdisciplinary thought. He was a trained physician who contributed knowledge to various academic fields, including physics, medicine, physiology, philology, and Egyptology. For the latter, he contributed valuable insight regarding the understanding of hieroglyphs. Standing out among the many professional positions he held are professor at the University of Cambridge, professor at the Royal Institution of Great Britain, and secretary of the Royal Society of London (institution that had previously awarded him with a scholarship at the age of 21). His curiosity and creativity led him to design famous experiments such as "the double slit," presented in 1801 at the Royal Society of London under the topic of "On the Theory of Light and Colors." This experiment was galvanized by his great fascination with light physics and was carried out with the inventiveness of a physicist (without actually being one). This elegant experiment characterizes light as a wave, which finally concluded with the long Newtonian era of the corpuscular theory of light. Young finished designing this experiment in 1803 with the text "Experiments and Calculations relative to Physical Optics." Here, he describes how light waves can be added or canceled among them. This experiment is currently essential in microscopy and optics to understand the nature of light. Consequently, an interdisciplinary center for the study of materials and created by various British universities, including Imperial College London, King's College London, Queen Mary University London, and University College London, bears his name: "The Thomas Young Centre"

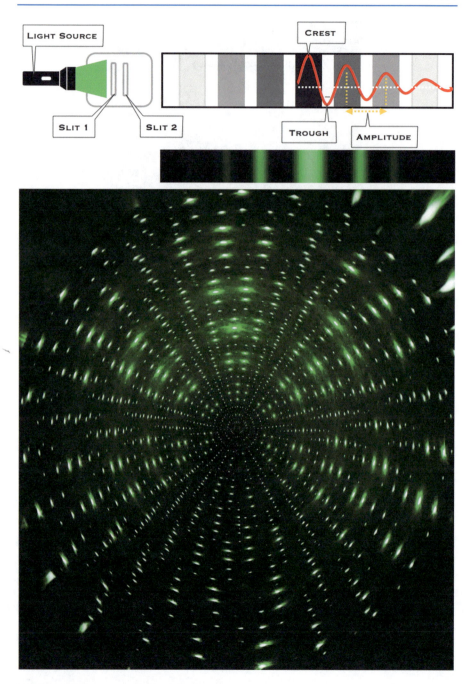

Fig. 2.3 Science, art, and light waves
Light is inherently chaotic, which means that when cast from a source, most photons will concentrate on a central axis, and the rest will rapidly scatter in multiple directions. Shown in the superior panel is a schematic diagram of an experiment made by Thomas Young, who projected light

2.2 The Duality of Light: Wave and Particle

Newton Prisms and Light interaction

Dispersion of white light through a crystal prism seen with LUT-fire filter of FIJI

To complement this section, it is important to mention that, for Huygens, the medium through which light waves move was an invisible medium, referred to as the luminous ether, a concept that was later disqualified due to the lack of evidence about its existence. James Clerk Maxwell (1831–1879 AC) verified that light travels through the electromagnetic spectrum and that it implies the interaction of light with solid, liquid, and gaseous media. This interrelation gives rise to different phenomena, including refraction, reflection, convergence, divergence, and dispersion.

2.2.1 Refraction of Light

Refraction consists of its passage through two media of different densities. If the passage of light from the first medium (n1) to the second (n2) is achieved at an angle of 90°, and if the interface of the medium does not contain topographic variations and it is totally perpendicular to the beam of light, the trajectory of the beam will not change when going from medium n1 to medium n2 (Fig. 2.5). However, if the light beam changes its inclination in medium n1, the path of the light will be affected in medium n2 (Fig. 2.6).

Fig. 2.3 (continued) through a double slit onto a screen, elegantly proving the undulatory nature of light. We can imagine light traveling in waves, forming crests and valleys. The distance between crests or between troughs is referred to as the amplitude. Each time a wave of photons passes through the first slit and meets another wave from the second slit, it will result in one out of two phenomena (regardless of its amplitude): either in the coordinated addition of waves of light (crests with crests and troughs with troughs) or in the cancellation of waves by the meeting of a crest and a trough. The first phenomenon will generate illuminated areas, brighter at their centers and dimmer at the extremes (due to the decrease in the number of photons, as illustrated by the gray bar scale in the upper panel and green bars in the middle panel), while the latter will stand out as areas of darkness (white areas in the upper panel and black areas in the middle panel). The double-slit experiment will always generate bar-like patterns of light, which are represented in this image as if seen through a computer imager. Can you tell how many double-slit images were required to make the image at the lower panel work?

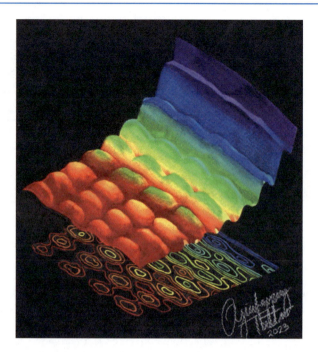

Fig. 2.4 Wave–particle duality seen under the transmission electron microscopy
A research group from the École Polytechnique Fédérale de Lausanne, Switzerland, led by Fabrizio Carbone, acquired photomicrographs of light, confirming in the process that light has a wave–particle nature, just as Albert Einstein had previously postulated in his theory of the photoelectric effect, which in 1921 earned him the Nobel Prize in Physics. What was the outstanding contribution of Carbone and his collaborators? Einstein postulated that light had a wave behavior on some occasions, and on others, it acted as a particle; however, there was no experiment where both conditions could be verified simultaneously. To achieve this, Carbone's group used a laser beam that was directed at a nanofilament. This filament was in contact on both flanks with an energy source, so it was charged with photons. When the laser was incident, it caused an overload of energy in the nanofilament, making it vibrate while it was producing light waves that, consequently, traveled in opposite directions until they met between themselves, generating in this way a wave that seemed to be in "pause" and that could be micrographed with a high-speed transmission electron microscope. The interaction between electrons emitted by the microscope on pulses of photons generated a wave–particle image similar to the watercolor replica shown in this figure
In this image, the progressive sinusoidal pattern reflects the stationarity of radiation, providing evidence of the wave nature of light. The "humps" (clearly visible in red) represent the energy quanta exchanged between electrons and photons during the experiment, demonstrating the particle-like behavior of light. This visualization simultaneously illustrates the dual nature of light. It is not derived from an actual optical micrograph but is instead a graphical representation of collected data. The two horizontal axes correspond to spatial dimensions, while the vertical axis (altitude) represents energy, which is higher for ultraviolet than for infrared, as expected

2.2.2 Reflection of Light

In contrast, the phenomenon of reflection does not imply the passage of light from an n1 to an n2 medium. In this case, light impinges on a surface at an angle of incidence that allows the interface of medium n2 to return the beam of light at an angle

2.2 The Duality of Light: Wave and Particle

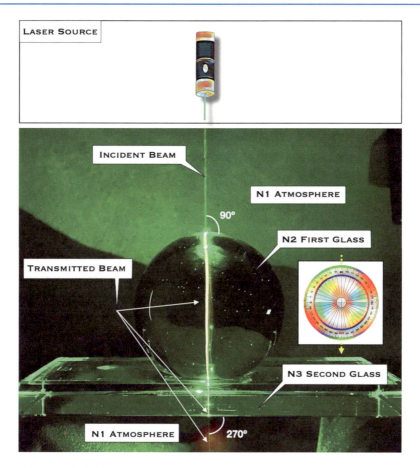

Fig. 2.5 Undisturbed light
It is possible for traveling light to interact with different types of media and to not be refracted, reflected, or dispersed. This provides that light travels through the given media in the 90° axis of the given media. In the image, we can observe how the beam of laser light leaving its source travels through the low-density atmosphere (n1 medium) toward a glass sphere (n2 medium, high-density medium) and hits the upper pole (entering at 90°). Because the point of incidence is right on the normal axis of the sphere, the light is not deflected and is transmitted along this vertical axis until it leaves the sphere at the lower pole (exiting at 270°, according to the multicolored graduated protractor) without deviating. It can even be appreciated how the light beam leaving the sphere passes through a new high-density medium, a second glass (n3), and returns to the medium n1 without variations on its vertical axis

of reflection within medium n1 (Fig. 2.6). It must be considered that reflection is possible even when the interface between the mediums n1 and n2 is irregular. If the beam of light interacts with an irregular n2 surface, it will be possible to produce reflection and refraction simultaneously.

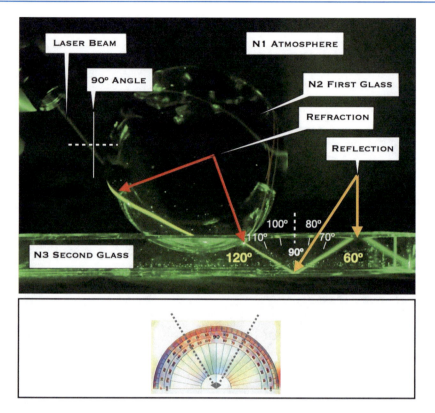

Fig. 2.6 Moving light
If a beam of light coming from an n1 medium falls at an axis different from the one corresponding to 90° at an n2 medium, the light will tend to change its direction of travel; this alteration is known as refraction. In the illustration, we see an oblique laser beam that passes from the atmosphere into a glass medium; in the latter, the deviation of the light can be observed. This same inclination, with respect to 90°, favors that when passing from medium n2 to n3 (second glass), refraction occurs again (red arrows). On the other hand, to ensure the reflection of a light beam, it must travel from a 60- to 120-degree angle. In this way, reflection will occur instead of refraction. In the photograph, two events of reflection are observed: the reflection occurs because a beam that was refracted from medium n2 to n3 has acquired an angle close to 120°, which when reflected by the upper edge of medium n3 acquires a new orientation, now with an angle close to 60°, which is reflected again (second event) within medium n3 (orange arrows). At no time does the light beam return to medium n1 from medium n3. Currently, various refractive indices have been described for different media, including water and oil, which are recurrently used in microscopy to correct the direction of light beams that are transmitted or reflected in a sample, with the purpose of improving the sharpness and resolution of the produced micrographs

2.2.3 Lens Geometry

Overall, if the light beam is oriented at a different angle to the normal one (which for simplicity will be considered 90°) when passing the n1 medium, it will tend to change its direction in the n2 medium (provided that the interface surface between n1 and n2 is always flat). It could also happen that the light beam moves from a 90° angle but remains parallel to an interface with a curved surface, this is the case for some lenses. In this hypothetical situation, there will be a change in the direction of the light beam, and the direction acquired in the n2 medium will depend on the type and degree of curvature of the lens surface and on its density. Regarding its geometry, there are lenses whose curvature moves away from its vertical axis. These are known as positive diopter lenses, convex lenses, or converging lenses, while those that come close to this axis of the lens are considered negative diopter lenses, also known as concave lenses or diverging lenses (Fig. 2.7).

2.2.4 Lens–Light Interaction (Convergence, Divergence, and Dispersion of Light)

Related to the light–lens interaction, while lenses with negative diopters will cause the light beams to disperse when passing through them, those with positive diopters will cause the light beams to concentrate on the horizontal axis (Fig. 2.8). In microscopy, knowledge of the light–lens interactions helps to understand the trajectory of the light beams to deliberately make them coincide at what is deemed the focal point. A correctly adjusted focal point favors the formation of magnified images that are sharp and have high resolution.

Circumzenithal arc: A miniature double turned upside down rainbow

This inversion is caused because light interacts with tiny ice crystals instead of water droplets.

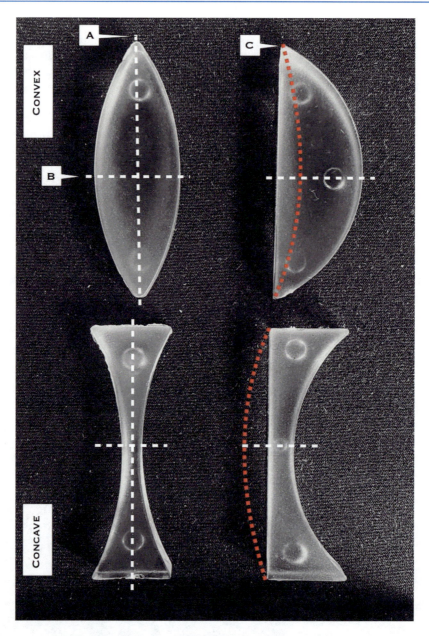

Fig. 2.7 The shape of the lenses: a resource to control the direction of light
Known as positive, converging, or convex, these lenses are characterized because their curvature is separated from their vertical axis (**a**), and the greatest convexity is reached at a point within the horizontal axis (**b**). The upper left lens in the image is double convex, which is also called double positive because positive numerical values (diopter) are assigned when the curvature moves further away from the vertical axis. The lens on the upper right side is referred to as plane convex, meaning that it contains a neutral diopter (flat side) and a positive diopter (curved side). The lenses on the

Another of the important light phenomena for microscopy is dispersion, which was first described in detail by Isaac Newton. He passed white light through glass prisms, achieving the decomposition of light into different wavelengths. This phenomenon of dispersion is common to find on a day-to-day basis. For example, when sun rays cross water or glass, they can create a rainbow, which is one of the most attractive manifestations of light dispersion and a pleasant image able to stimulate the minds of all observers, regardless of age, inviting vivid questions and explanations that range from the magical to the scientific (Fig. 2.9).

In the field of microscopy, knowledge of the different wavelengths of light has contributed to the generation of light filters that can recover the colorimetric information emanating from any wavelength. This is useful, for instance, when introducing fluorescent-colored labels in biological samples attached to molecules of interest in tissues and cells.

2.2.5 Lens Aberrations

However, when applying these techniques, not all the obtained dispersion is useful for our means. For instance, the ends of a converging lens are practically prisms. In these areas, the light breaks down, resulting in alterations in the color of the final image, which is called chromatic aberration. Moreover, lenses, given their imperfect shape and finish, can produce other visual aberrations, such as astigmatism and spherical aberration, both associated with the displacement of the focal point in the horizontal axis of the lenses, although the latter is associated with highly positive diopter lenses, while astigmatism is associated with lenses that do not have uniform finishes, producing different focal points. Another aberration is produced by the misalignment of light and is called comatic aberration or coma aberration, which resembles the shape of a comma or a comet, hence the name. Another common aberration is called field curvature aberration, which appears as an out-of-focus curvature at the edges of an image projection, a focused area along the axial axis. It partially disrupts sample focus because it is possible to adjust the focus of the edges, though as a consequence results in loss of focus at the axial region (Fig. 2.10).

Fig. 2.7 (continued) lower side of the image are negative diopter lenses, also called divergent or concave lenses. The lower left lens is a double-negative diopter lens, unlike the positive lenses; here, the curvature seems to approach the vertical axis. Finally, the lower right-side lens is a flat-concave lens with one flat side. In addition to the lenses pictured, it is important to mention the meniscus-convergent and meniscus-concave lenses. The former has a curvature instead of a flat side that goes beyond the vertical axis and has been assigned a negative value, which is represented by the red dotted line (**c**) in the upper right image. These lenses are also called concave–convex lenses because they contain the combination of both curvatures and can also be called negative–positive lenses. Finally, represented by the dotted red line in the lower right image is a hypothetical concave–convex or positive–negative lens

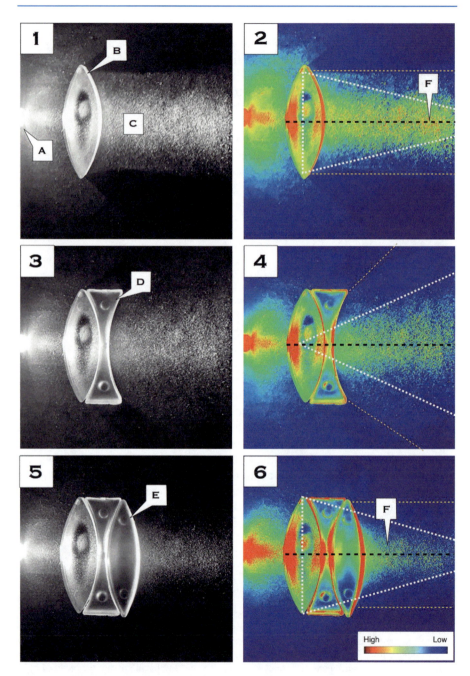

Fig. 2.8 Scattering and Concentration: Light–lens interaction exposed through a digital filter Panel 1: We can observe that when light emitted by a source (**a**) passes through a converging lens (**b**), the resulting rays seem deformed, with most of the light located within a central area (**c**). This reduction of the illuminated area is due to the converging lens, which causes a concentration of the traveling light. Panel 2: Here, the same image from the first panel is shown with an innovative

2.2 The Duality of Light: Wave and Particle 27

Fig. 2.8 (continued) lookup table (LUT) digital filter applied. This filter employs color tones to distinguish different energy levels, with red meaning greater energy, blue meaning little to no energy, and orange, yellow, green, and cyan tones representing intermediate energies (see scale bar in 6). The filter reveals that the greatest amount of energy (thus the highest accumulation of photons) is found mainly when the light emerges from the lamp and collides with the surface of the lens, causing it to be refracted or reflected, according to its angle. After passing through the lens, the light travels in a cone-like formation (highlighted with dotted white lines) made up of red, yellow, and green areas. This shape of dispersion is due to the lens shape, which will promote photon concentration toward a central horizontal axis, known as the focus or focal plane (**f**). The blue area outside the cone, highlighted with the perpendicular orange dotted lines, shows scarce to no presence of energy. Panel 3: This image proves that by placing a diverging lens (**d**), the light that would previously be concentrated will now be dispersed from the horizontal axis. Panel 4: The added diverging lens causes the light dispersion to invert shape (white dotted lines), the focus of 2 to disappear, and the blue areas toward the center of image 2 to decrease. Panel 5: Finally, this figure shows that to reconcentrate the light scattered by the concave lens back to the previous focus, the placement of a second convex lens (**e**) is needed. Panel 6: A second convex lens will again result in the convergence effect and, prominently, in the reduction in the focal plane length

Fig. 2.9 Double rainbow at the Soleil synchrotron building

2.2 The Duality of Light: Wave and Particle

Fig. 2.9 (continued) Rainbows are meteorological events generated by the interaction of light with raindrops. Depending on their angle when the light beams enter a drop, some will be refracted, or reflected, or others will have their wavelengths dispersed. Subsequently, they will reach the innermost face of the drop; here the light will either (1) exit from the drop after refraction or (2) get out of the drop in the opposite direction after being reflected (above, left). Both events will also be dependent on the angle at which the face of the drop receives the light. In the case of reflection, the light will be dispersed into the drop, return and exit through the face where it enters, projecting itself onto the horizon like a rainbow that will generally retain the same staggered arrangement of the colors that compose it, red on the outside, followed by orange, yellow, green, blue, and ultraviolet (above, right). Infrared light contains a long wavelength (low energy) that makes it difficult to deflect, while its opposite, ultraviolet light, has shorter wavelengths (high energy) that allow for easier deflection (second panel). This results in an arc of greater curvature that is located in the inner part of the rainbow. Not all rainbows have the same arrangement. In the photograph, you can see a second rainbow (an early description was made a long time ago in 1878 in *Nature* journal) that is inverted and positioned on top of the primary rainbow. This is because the first reflection of light inside the drop has a different angle and insufficient energy for the light to return and leave the drop. Given its angle, a second reflection occurs, leading to the inversion of the colors and producing a lower intensity (below). Strange and fortunate phenomenon for those who observe it, the angle from which this photo was taken makes it seem like both rainbows are emerging from the Soleil Synchrotron facility building (at Saint-Aubin, France). Synchrotron is a powerful microscope that uses highly accelerated electrons to understand the chemical composition of samples at low and high resolution

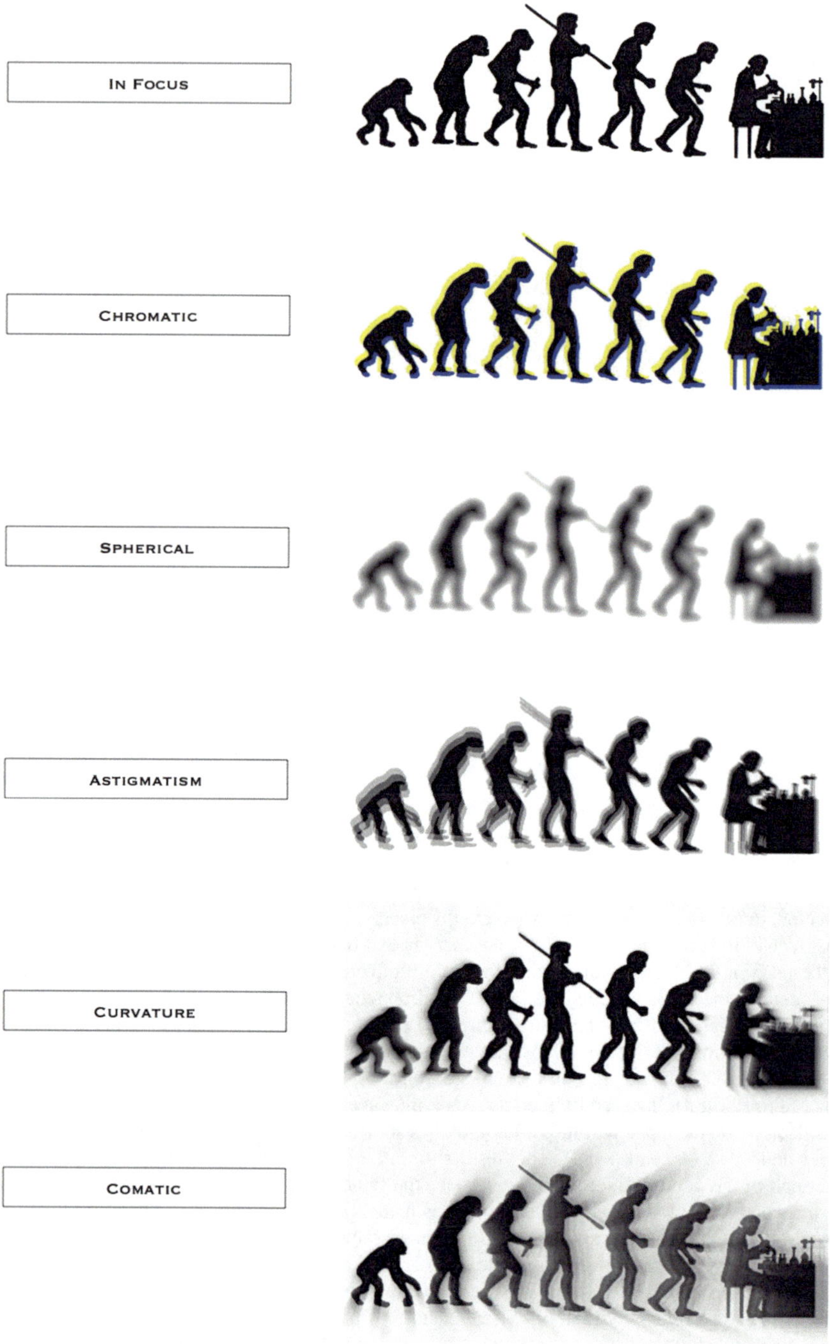

Fig. 2.10 Aberrations

Fig. 2.10 (continued) *Chromatic*: A color distortion observed at the edges of the sample. It is caused by focal displacement along the optical axis due to the different wavelengths of white light-*Spherical*: A change in focal points within the image plane due to light interacting with a lens with curvatures that extend farther from their axial path. High-curvature lenses are more likely to exhibit this aberration

Astigmatism: An imaging defect where different focal planes form along the axial axis due to lens shape and inherent deformations

Curvature: Is an out-of-focus curvature at the edges of an image projection, with a focused area along the axial axis. It partially disrupts sample focus, as adjusting the focus at the edges results in loss of focus at the axial region

Comatic: It occurs when light refracts through lenses at increasing angles of incidence or due to improper alignment of lenses or light sources. This aberration creates asymmetrical images that gradually increase in size, resembling the shape of a comet

Further Reading

For ligth and lenses: Batchelor (2012), Piazza et al. (2015), Sliney (2016).

For rainbow: Corradi (2016), Naylor (2023), Noyé (1878).

For double-slit experiment: Crease (2002), Rosa (2012). Nobel Prize (2022) quantum theory.

For Einstein and Newton: Simonton (2013), and Storr (1985).

References

Batchelor BG (2012) Light and optics. In: Batchelor BG (ed) Machine vision handbook. Springer, London. https://doi.org/10.1007/978-1-84996-169-1_5

Corradi M (2016) A short history of the rainbow. Lett Mat Int 4:49–57. https://doi.org/10.1007/s40329-016-0127-3

Crease RP (2002) The most beautiful experiment. Phys World 15:19–20

Naylor J (2023) The riddle of the rainbow. From early legends and symbolism to the secrets of light and colour. Copernicus Books, Springer, Cham. https://doi.org/10.1007/978-3-031-23908-3

Nobel Prize (2022) MLA style: the Nobel Prize in Physics 1921. NobelPrize.org. Nobel Prize Outreach AB 2024. https://www.nobelprize.org/prizes/physics/1921/summary/

Noyé TA (1878) Double rainbow. Nature 17:262. https://doi.org/10.1038/017262a0

Piazza L, Lummen TTA, Quiñonez E, Murooka Y, Reed BW, Barwick B, Carbone F (2015) Simultaneous observation of the quantization and the interference pattern of a plasmonic near-field. Nat Commun 6:6407. https://doi.org/10.1038/ncomms7407

Rosa R (2012) The Merli-Missiroli-Pozzi two-slit electron-interference experiment. Phys Perspect 14(2):178–195. https://doi.org/10.1007/s00016-011-0079-0

Simonton D (2013) Scientific genius is extinct. Nature 493:602. https://doi.org/10.1038/493602a

Sliney DH (2016) What is light? The visible spectrum and beyond. Eye (Lond) 30(2):222–229. https://doi.org/10.1038/eye.2015.252

Storr A (1985) Isaac Newton. Br Med J (Clin Res Ed) 291(6511):1779–1784. https://doi.org/10.1136/bmj.291.6511.1779

The Brightfield Microscope

3

By the help of microscopes, there is nothing so small, as to escape our inquiry; hence there is a new visible world discovered to the understanding.

Robert Hooke

© The Author(s), under exclusive license to Springer Nature Switzerland AG 2025
A. Rosas-Arellano et al., *Microscopic Wonders*,
https://doi.org/10.1007/978-3-031-92559-7_3

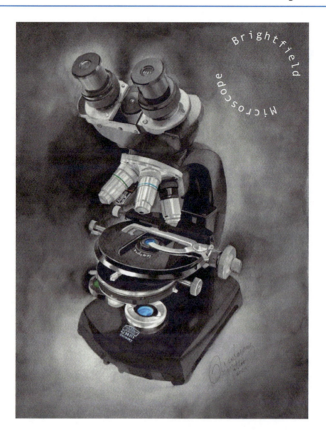

As it is inherent in microscopes, the brightfield specimen is composed of three main systems: (1) optical, (2) illumination, and (3) mechanical (Fig. 3.1). Brightfield microscopy, also called white-light microscopy, gets its name from the tone of the lighting it uses, which can be of three types: (1) incandescent, (2) halogen, or (3) light-emitting diode (LED).

A microscope will always be designed to optimize the performance of its lighting resource (Fig. 3.2). Without downplaying the importance of the other systems, lighting is of utmost importance, since it will interact with the sample and return to the observer the information necessary to analyze it. In microscopy, recovering the largest amount possible of the light that passes through a sample is regarding the resolution and will avoid loss of information and a consequent poor interpretation of results. When light interacts with a sample, there will be many fascinating phenomena taking place, with two of the most important being light absorption and transmission. In the case of absorption, some photons will be incapable of passing through the sample, generating regions that will appear colored in darker tones.

On the other hand, transmission occurs when photons have little to scarcely obstruction when passing through a sample. This will appear to the observer as less

3 The Brightfield Microscope

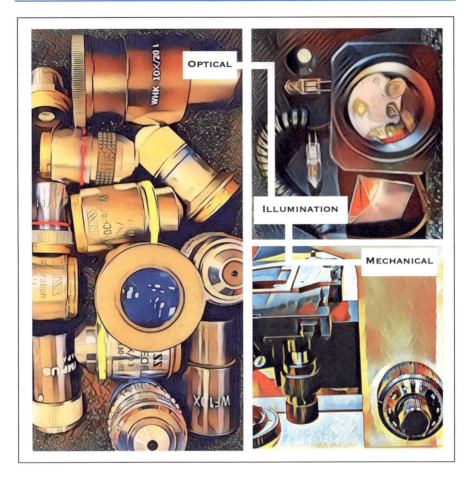

Fig. 3.1 Some interconnected characters in cartoon form
Optical system: Composed of the collector lens, field lens, condenser lenses, objective lenses, aberration correctors, eyepiece, and prisms, among other lenses*Illumination system*: It consists strictly of the lamp (illuminator), which is the lightning source; although by convention it could also include mirrors, or polarized filters, as well as the light intensity regulator (brightness adjustment)
Mechanical system: The most complex of all features, made up of the condenser diaphragm (iris), adjustment screws (including those associated with the condenser's focus), coarse focus knob, fine focus knob, stage, sample holder clamps (stage clips), arm, revolving nosepiece, and eyepiece tube, among other parts

dense and more brightly colored areas. It is because of this phenomenon that this type of microscopy is also referred to by the name transmitted light microscopy. Consequently, it is because of absorption and transmission that brightfield photomicrographs will be characterized by containing both dark and light areas (Fig. 3.3).

Other consequences of the light–sample interaction in brightfield microscopy, beyond absorption and transmission, include phenomena that provide us with useful

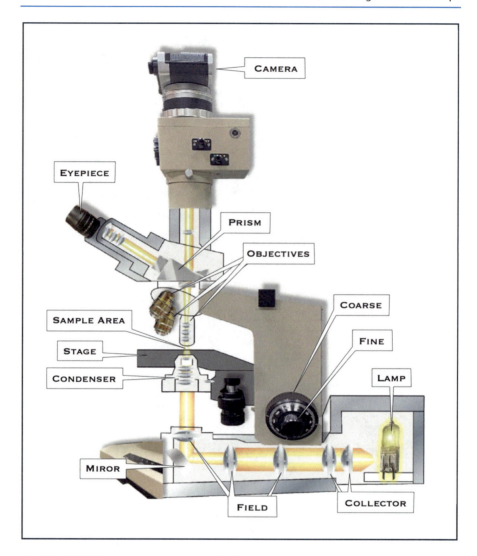

Fig. 3.2 Brightfield microscope (compound light microscope)
Depending on the microscope, the path followed by the light will vary. In this case, the light from the source is captured and alienated by the collector lenses. The first of these lenses has a flat–convex (neutral–positive) finish, allowing recovery of the greatest amount of light through its flat side, while the convex side concentrates the light into a beam that is directed into the second collector lens, which has a double convex shape. The latter lens prevents the light from diverging and directs it toward a set of field lenses also of a convergent nature; thus, the beam extends aligned to a mirror that redirects the light toward the condenser lens set

The function of the condenser lens set is to concentrate the light beams coming from the collecting lenses into the sample placed on the stage. The light will then be directed to the objective front lens, a convex-type lens that aligns the light toward the rest of the lenses located inside the objective. The objective lenses, in addition to recovering the light information coming from the sample positioned in the stage, will simultaneously magnify the sample's image and are the first magnifying lenses

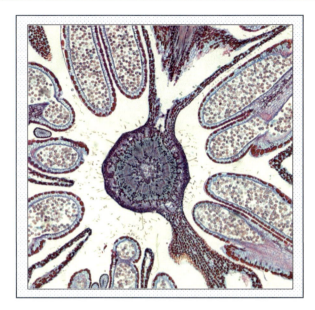

Fig. 3.3 The colors of absorption and transmission
A biological sample intended for observation under a brightfield microscope can be stained with dyes to show high contrast between the dense and non-dense structures that compose it. Stained samples are typically nonliving specimens that may have been fixed with aldehydes to adequately preserve their microscopic structures. This type of staining, also called colorimetric, specifically shows the presence of selected biomolecules such as carbohydrates, proteins, or nucleic acids. This photomicrograph shows a histological section of *Corylus Hasel* stained with triple Masson stain (with dyes: fibers, nuclei, and cytoplasm) and taken with AxioZoom.V16, a specialty stereoscopic microscope

selective information (Fig. 3.4). For example, surface information can be obtained by reflection microscopy; in this type of microscopy, transmission and absorption are not relevant. There are various types of reflection microscopy as total internal reflection fluorescence (TIRF) microscopy and scanning electron microscopy (SEM); both of them will be reviewed later.

◄

Fig. 3.2 (continued) There usually is a set of lenses between the objectives and the eyepiece lenses whose purpose is to infinity-correct the image. This means that distortion of the image passing from the first objective lens to the eyepiece lens is prevented. Prior to reaching the ocular lenses, the transmitted beam of light will pass through an optical prism that helps to invert the image so that the X and Y axes of the stage are synchronized with the spatial orientation of the observer. This allows for the specimen to be manipulated without having to manually invert the upper, bottom, left, or right axesFinally, the information will reach the eyepiece lenses to be directly observed through the eyepiece or captured by a camera for computational analysis

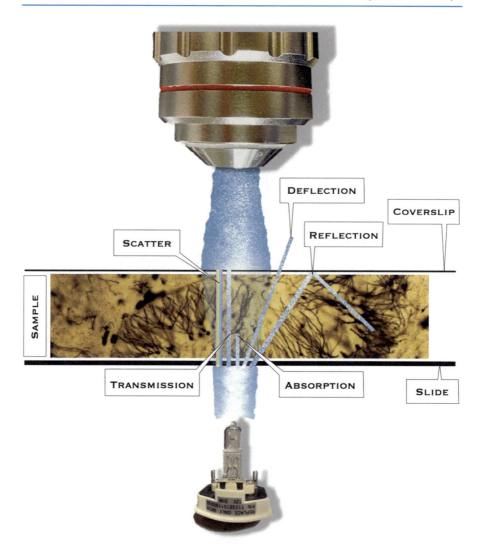

Fig. 3.4 More than absorption and transmission of light
Light absorption and transmission are the most important events for brightfield microscopy. However, as illustrated in the figure, light generates other types of major interactions with the sample; this is the case for absorption, reflection, refraction, dispersion, convergence, and divergence. Although each of these events involves an inherent loss of light information, they are still considered cornerstones of microscopy since each of them is key in achieving all types of microscopies that we know today

The neurons fossilized in amber

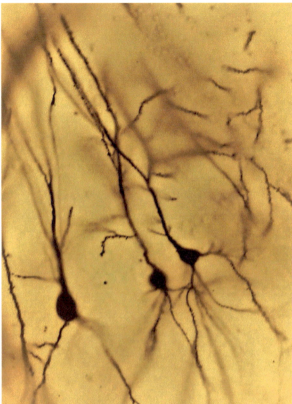

Pyramidal cells in the hippocampus, Golgi stain

In transmitted light microscopy, the light that has passed through a sample is directed toward the microscope's objectives, which will contain arrangements of inner lenses with various purposes. For instance, objectives are the first magnifying lens, and they can correct aberrations, filter the light, or be coupled with immersion media to promote better resolution. There are endless ways to fashion objective lenses to make proper use of them. When purchasing them, these specifications will be engraved on their sides, alongside colored lines that are used to identify their magnification and any immersion media they could be designed for (Fig. 3.5). Related to magnification, color-code was designed to be a representation of the energy of the visible spectrum according to the resolving power of the objective lenses (Fig. 3.6).

Similarly to objective lenses, ocular lenses have specialized optical configurations to eliminate aberrations and to adapt to the eyesight of the observer (Fig. 3.7). Additionally, they can even have anti-fungal or anti-reflective coatings. The eyepieces, however, are only used for observing and properly framing a region of interest that can be subsequently captured through a camera adapted to the microscope;

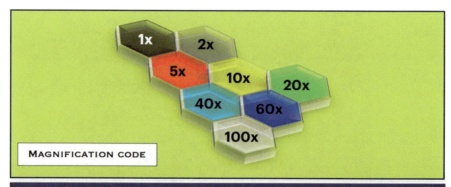
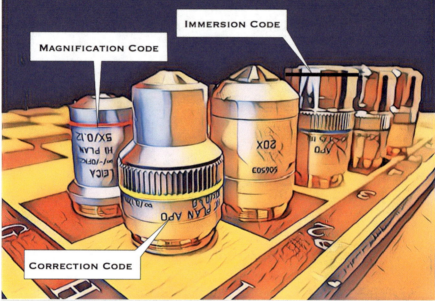
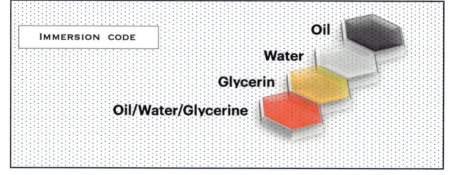

Fig. 3.5 Objective strategy
The objectives of a microscope are designed in a highly selective manner. Even if we had two objectives with the same magnification, they may not be exactly the same and could vary in their corrections, light filters, numerical aperture, and immersion conditions. To distinguish their characteristics, each objective contains information engraved instructing on their correct use.

3 The Brightfield Microscope 41

therefore, it is better to invest in a high-resolution camera than in powerful eyepiece lenses.

Lesser known is the iris diaphragm, a metallic assembly that can be adjusted to limit the amount of light entering the objectives and condenser. Importantly, if we are required to employ high magnifications to properly observe our area of interest in a sample, then the condenser diaphragm must be kept at smaller apertures with a slight increase in light intensity. Increasing the closure of the diaphragm favors resolution by facilitating and concentrating the arrival of optimal light concentrations to the micrometric area of interest. On the other hand, when the employed magnification is low, then the condenser diaphragm opening must be larger to favor ample illumination of all the observed area and thus recover the greatest amount of information (Fig. 3.8 top).

As previously introduced, proper lighting is highly important when it comes to making correct use of a brightfield microscope. It is thus essential to make sure the light beam employed remains focused on the area of interest. For this reason, a mechanical resource to adjust illumination has been adapted into microscopes. It is known as Köhler lighting (Fig. 3.8 bottom).

However, to ensure proper harnessing of a microscope's potential we must also make sure to keep our lenses' system free of dust and grease. When present, these particles will not only alter the information obtained from a sample, but they can also generate colorimetric patterns that can be erroneously attributed to the specimen observed (Fig. 3.9). Therefore, periodic but gentle cleaning of the lenses is essential, in addition to keeping the microscope properly covered when not in use. An extremely efficient binary solution for removing dust and grease is made from absolute ethanol and acetone in a ratio of 9:1, respectively.

There are yet several other components counted as essential when it comes to employing microscopes in an optimal manner. Among them is the distance kept between the upper edge of the sample and the outer edge of the objective's front lens. This distance is called working distance (Fig. 3.10 top). Within this same area, between the sample and the objective front lens, there is an important element that directly influences the resolution: the numerical aperture. For each objective, this value will be engraved next to the corresponding magnification (Fig. 3.10 bottom).

When light passes through a sample and encounters air it will inevitably be widely diffracted. As a convention, the air medium has been assigned a refractive index of 1. If we strive for the highest resolution possible, we must attempt to recover the largest number of light beams possible. Due to air interactions, another media with a higher refractive index (that is, with higher density than air) will have to be used; this is when immersion media comes into play (Fig. 3.11). Commonly

---

Fig. 3.2 (continued) For instance, the plan apochromat (Plan Apo) engravings mean the objectives were designed to correct spherical and chromatic optical aberrations. Also included in the engraved information are indications of numerical aperture, thickness of coverslips that allow adequate transmission of light, and working distance for adequate focus; magnification is also indicated by a color band just below the rest of the information (upper panel), and sometimes when immersion is used, there is also another color band indicating immersion medium, just under the magnification band, close to the front lens (lower panel). As you can imagine, picking the correct set of objectives to reach optimal resolution conditions for each sample becomes a true chess tactic

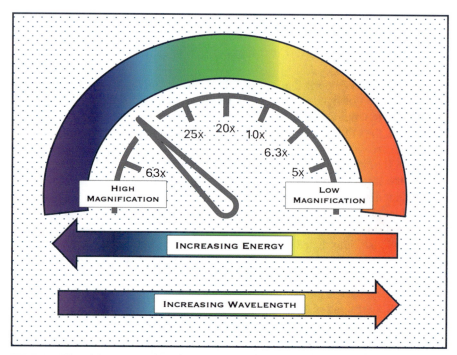

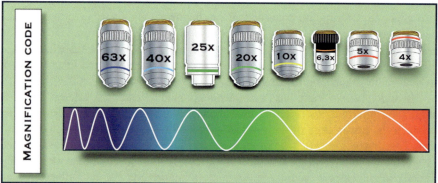

Fig. 3.6 Encoded magnifying power

The power of objective lenses has been color-coded according to the energy of the color wavelengths of the visible spectrum; this association makes it easy to associate whether a color band corresponds to an objective with higher or lower magnification. As mentioned before, the highest magnification power objectives are colored for high energy wavelengths (blue), the intermediate ones are color-coded with green, yellow, and orange, while the lowest magnification power objectives correspond to low energy tone (red). At the extremes of this spectrum are the black and white codes indicating magnifications near or corresponding to 1× and 100×, respectively, while 3× and 2× are usually represented with brown colors

3 The Brightfield Microscope

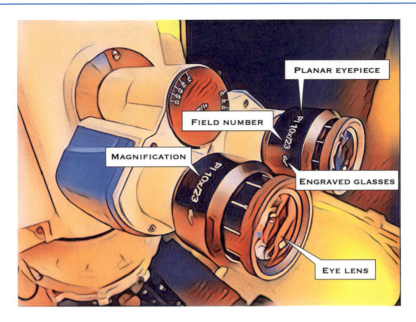

Fig. 3.7 Eyepiece anatomy
Each of the eyepieces culminates in a lens known as an eye lens and is usually the upper element of a microscope. The sample area that can be covered by the eye lens is called the field of view, which is provided by the diameter of the lens between the magnification of the employed objective. The diameter in millimeters of each eyepiece is known as the field number and can be found engraved on its side. With this information, we can infer that, for instance, an eyepiece with a field number of 20 millimeters and a 40x objective will possess a field of view of 500 micrometers. Knowing this information becomes relevant when making density estimates. Eyepieces will also be engraved with more information including their magnification and correction. The latter means, for instance, that someone with eyesight impairments can use the eyepieces without having to take their glasses off (engraved with little glasses). Also relevant is that some eyepieces have a widefield (engraved as WF), which means that the lenses will cover an area 50% larger than conventional eyepieces. Moreover, eyepieces can have flat field (engraved as Pl) eye lenses, which will not visually distort the observed area

used media are water and oil; since their refractive indexes are close to those of glass, they will significantly limit diffraction and will function as flexible lenses that can aid in re-orienting the angles at which light beams are traveling, directing them toward the area of the objective lens.

The numerical aperture and the wavelength used to illuminate a sample are important data to calculate the resolution of a microscope; this will be reviewed in detail in Chap. 7.

Related to energy, white light is made up of a wavelength between 400 and 500 nm. Together with a numerical aperture of 1.0 or 1.5, we can obtain a resolution of around 400 nm for a brightfield microscope under optimal equipment, sample, and ambient conditions. This leads us to two more concepts: the limit and the resolution power of a microscope, both of which are detailed in Chap. 8.

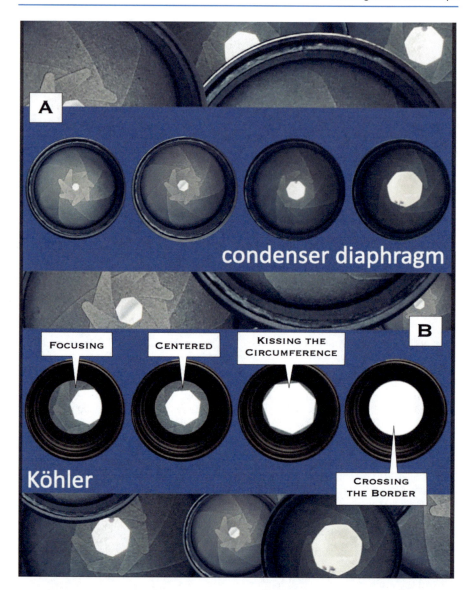

Fig. 3.8 Harmonic visual framing. (**a**) The condenser diaphragm is a dynamic element that is useful for adjusting illumination when changing the objective magnification and for aligning the light beam in order to achieve optimal sample illumination. (**b**) Created by August Köhler, Köhler illumination allows correction of misaligned, uneven, and overall poor sample illumination, resulting in images with improved sharpness and resolution. With this adaptation, illumination is adjusted in out-of-focus planes, which prevents the light source itself from generating artifacts during specimen observation or capture. Adjustment is achieved by employing various elements that compose the condenser, including diaphragm, focus knob, XY condenser adjustment knobs, and an objective with low magnification

Once the diaphragm is located and focused in its characteristic seven-sided design using the condenser focus screw and looking through eyepieces (focusing), an attempt is made to center it in

3 The Brightfield Microscope

Fig. 3.9 Whimsical forms of dustTransmitted light and accumulated dust particles on the condenser lens have resulted in diffracted light forming these strange and unexpected colorimetric arrangements. These types of occurrences could compromise result obtention and interpretation. It is thus essential to perform throughout cleaning of our microscopes in a periodic fashion

Brightfield microscope can be purchased in various models including the upright, inverted, and stereoscope (Fig. 3.12). These can be further adapted to specific means with illumination filters or light-obstructing adaptations; these have the power to highlight different aspects of the sample and are especially relevant when it comes to working with *in vivo* samples, which cannot be stained with dyes. Some illumination adaptations include differential interference contrast, phase contrast, and dark field (Fig. 3.13).

Fig. 3.8 (continued) reference to the field of view by adjusting the X and Y condenser screws (centered). Subsequently, the iris diaphragm is opened just enough so that each of the vertices of the polygon is barely reaching the circumference of the field of view (kissing the circumference). To align the sample illumination, we will continue to slowly open the diaphragm until the polygon has almost disappeared from the field of view (crossing the border). Subsequently, it will be necessary to adjust the focus of the condenser to make it coincide with the focus of the sample. This type of illumination must be periodically adjusted and is useful for all variants of brightfield microscopy equipped with a condenser

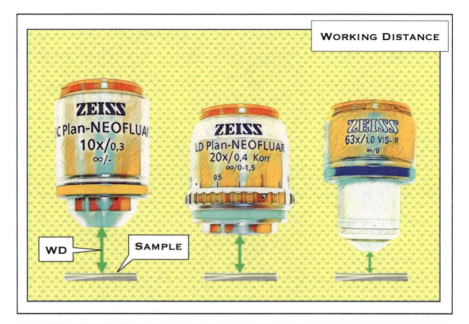

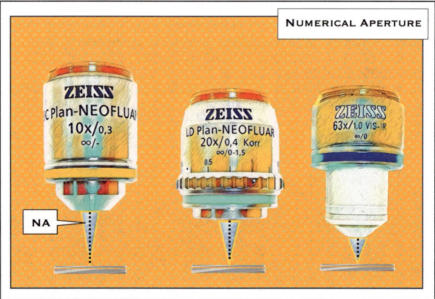

Fig. 3.10 The logistics between the objective and the sample
Upper: In microscopy, the working distance (WD) is known as the distance between the external face of the objective front lens and the upper face of the sample preparation (which is usually the upper face of the coverslip). This measure corresponds as long as an adequate focus has been achieved. As the image suggests, the working distance has to be reduced as the magnification of the objective increases. Consequently, a 10× objective will have a greater working distance than a 63× objectiveLower: In an objective, the numerical aperture (NA) is the value located on the right

Talavera cerebellar crafts

Purkinje cells in the cerebellum, toluidnine blue stain

3.1 When Should a Brightfield Microscope Be Used?

In its conventional configuration (transmitted light), it is commonly used to observe tissues, cells, or microorganisms that have been previously stained with natural pigments that generate high contrast in the samples. Some examples include hematoxylin-eosin and Golgi staining, as well as protein localization techniques like immunoenzyme (widely known as immunohistochemistry).

When the microscope is equipped with dark field, phase contrast, polarized light, or differential interference contrast, it allows for the observation of live samples without the need to stain them. This is beneficial as staining can sometimes lead to toxicity-induced damage or cell death.

Finally, it is suitable for observing molecules of interest that exceed the resolution limit, which is approximately 400 nm. This is providing that the microscope has the appropriate objectives, adjustment, and alignment; that the sample slide and coverslip are free of grease and dust; and that it has a thickness of between 20 and 30 micrometers.

Fig. 3.10 (continued) side of the slash, next to the magnification. Its value increases as the magnification does, and it is critical to consider it if striving for efficient microscope use. It corresponds to the angle at which the light enters the objective. As you can imagine, larger numerical apertures are the most sought-after, as they will allow for larger amounts of light, and thus information, to enter the objective, resulting in higher image resolution

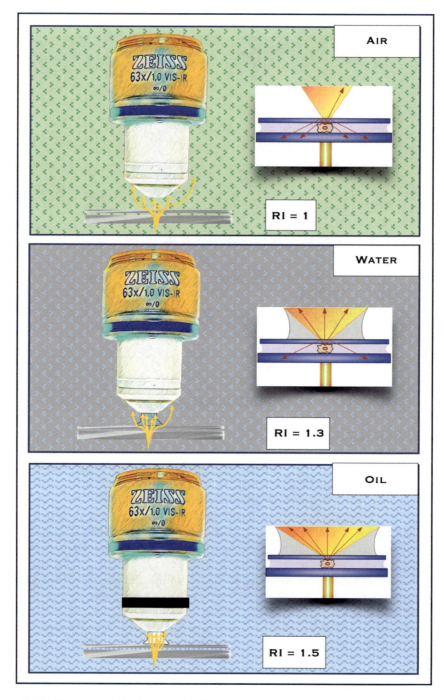

Fig. 3.11 The irreplaceable function of the immersion medium

3.1 When Should a Brightfield Microscope Be Used?

Fig. 3.11 (contined) Immersion media are used to recover light information from the sample. Air has a refractive index (RI) value of 1 and causes a decrease in resolution and loss of information due to an abrupt change in the direction of light rays that pass through the specimen preparation and encounter a low-density medium. However, water, with a refractive index of 1.3, promotes information recovery due to its higher density and by decreasing diffraction of the light passing through the sample. Oil media can further avoid loss of information due to its refractive index, which is practically the same as glass, the same material that coverslips and lenses are fabricated with. This interface glass–oil–glass of density homogeneity prevents light beams from diffracting and favors high resolution. In summary, the resolution of a sample is dependent on the characteristics of the wavelength, the numerical aperture, and the refracting index of the medium the light beams encounter when leaving the sample. Thus, to achieve maximum resolution when using a multi-immersion objective (for example, those suitable for both water and oil) and it is necessary to observe a sample that is not immersed in a liquid medium (such as culture medium), immersion oil should always be used to promote a better resolution

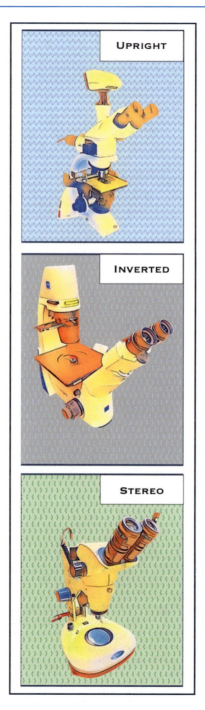

Fig. 3.12 Upright microscope

Fig. 3.12 (continued) With an inbuilt lighting system and a condenser below the stage, it is commonly used for the visualization of samples fixed with aldehydes or other fixative solutions and stained with various enzymatic dyes. It is one of the most widely used microscopes in laboratories

Inverted microscope

This type of microscope has its source of illumination, and the condenser is placed above the stage and the objectives below, hence its name. Its use is very popular when it comes to live sample observation, such as cells or tissues, which should not be stained. To optimize their observation, the microscope will usually count with a system of filters such as differential interference contrast (DIC), phase contrast, and dark field, among others

Stereo microscope

This microscope has a wide working distance (cm), a single objective, and two types of illumination: epi-illumination and trans-illumination (upper and lower). This microscope is conventionally employed for observing large samples; the enormous space between the stage and the objective lens is ideal to perform dissections or other elaborate manipulations

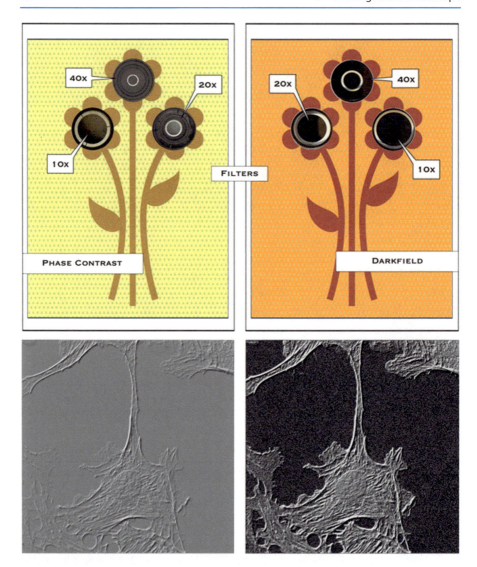

Fig. 3.13 Obstructed light
To obtain more complete information about a sample than what can be obtained from brightfield microscopy, especially for those specimens that cannot be stained due to the toxic effects of dyes in living organisms, diverse adaptations can be employed. These are polarized filters or dark circles that prevent the passage of the light beam's central axis, allowing us to observe the contours of the specimen that is being illuminated by the beam's lateral axis. The polarizer versions of these filters, which come in various colors, can also indirectly make a sample appear dyed. Light filters and light-obstructing adaptations are manufactured in several forms they can be included between the light source and the condenser, or even inside the objectives. The latter can be recognized by an auxiliary filter identifier located on its lateral side. For instance, objectives with adapted polarized filters will have the prefix Pol engraved; objectives and condenser for those including obstructed light adaptation as phase contrast will be Ph, DF for darkfield, and DIC in the case of those that have differential interference contrast filters. The resulting micrographs can be seen at the bottom for Ph and darkfield

Further Reading

Microscopy basis: Barrera-Escorcia H. and Cardenas-Reygadas (1997), Rochow & Tucker (1994).

For phase contrast and interference: Barer (1966).

For light microscopy: Chiarini-Garcia & Melo (2010), Käthner & Zölffel (2016), Lawlor (2019), Nechyporuk-Zloy (2022), Richardson (1991).

For DIC microscopy: Shribak & Inoué (2008).

References

Barer R (1966) Phase contrast and interference microscopy in cytology. In: Physical techniques in biological research, vol 3, pp 29–90

Barrera-Escorcia H., Cardenas-Reygadas (1997) El microscopio óptico. ISBN: 9688564818 (Plaza y Valdes) 968-36-6017-7 (UNAM)

Chiarini-Garcia H, Melo RCN (2010) Light microscopy: methods and protocols. eBook

Käthner R, Zölffel M (2016) Light microscopy. Mi and ZEISS

Lawlor D (2019) Introduction to light microscopy: tips and tricks for beginners. Springer, Cham. https://doi.org/10.1007/978-3-030-05393-2

Nechyporuk-Zloy V (2022) Principles of light microscopy: from basic to advanced, vol VIII. Springer, Cham, p 324

Richardson JH (1991) Handbook for the light microscope: a user's guide

Rochow TG, Tucker PA (1994) A brief history of microscopy. In: Introduction to microscopy by means of light, electrons, X rays, or acoustics. Springer, New York, pp 1–21. https://doi.org/10.1007/978-1-4899-1513-9_1

Shribak M, Inoué S (2008) Orientation-independent differential interference contrast microscopy. In: Collected works of Shinya Inoue: microscopes, living cells, and dynamic molecules, pp 953–962. https://doi.org/10.1142/9789812790866_0074

The Widefield Microscope

4

In 2008, the Nobel Prize in Chemistry was awarded for work done on a molecule called green fluorescent protein that was isolated from the bioluminescent chemistry of a jellyfish, and it's been equated to the invention of the microscope in terms of the impact that it has had on cell biology and genetic engineering.

Edith Widder

© The Author(s), under exclusive license to Springer Nature Switzerland AG 2025
A. Rosas-Arellano et al., *Microscopic Wonders*,
https://doi.org/10.1007/978-3-031-92559-7_4

The colorful journey of Aequorea victoria

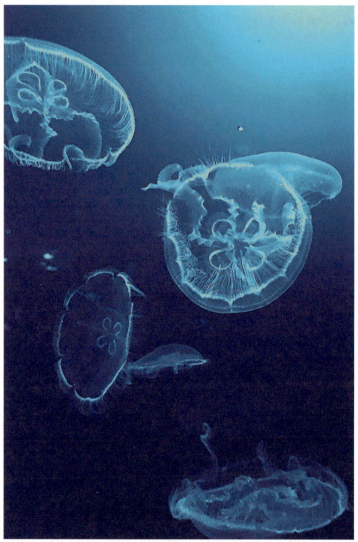

4 The Widefield Microscope

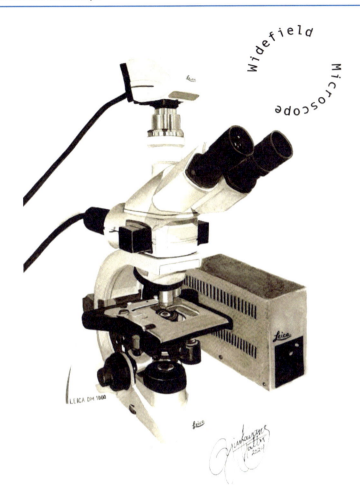

A wide range of microscopes belong to the fluorescence microscopy family, including (a) widefield, (b) confocal, and (c) super-resolution. This family specializes in the recovery of fluorescent light emitted by molecules contained in biological samples, known as fluorophores, the active element of a fluorochrome.

Although often employed as synonyms, the terms fluorochrome, fluorophore, and chromophore are, strictly speaking, not the same. A fluorochrome is a molecule that absorbs energy; on the other hand, when we talk about a fluorophore, we are referring to the functional group in fluorochromes that is responsible for the fluorescent emission that follows energy absorption. Finally, a chromophore is the functional part of fluorochrome complexes that determines the color of the emitted fluorescence.

Some organisms, such as bioluminescent plankton, can produce fluorochromes naturally. This is known as primary fluorescence or autofluorescence. Nevertheless, fluorochromes can also be designed, synthesized, and artificially introduced into biological samples through genetic engineering. This is known as secondary fluorescence.

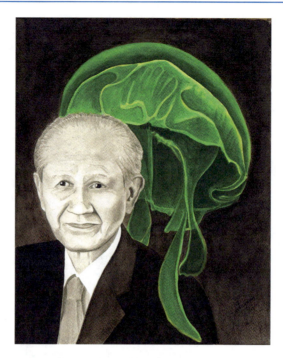

Fig. 4.1 Let there be light, and there was light

Osamu Shimomura, 1928–2018, Japan. The founder of one of the most momentous scientific tools in the history of modern science: the isolation of green fluorescent protein (GFP) from the jellyfish *Aequorea Victoria*. Since this protein was structurally described and cloned to be expressed in a fused state to biological molecules within different living organisms, thousands of behavioral, physiological, and pathological processes have been uncovered. Thanks to the fluorescent green light emitted by this protein, a wide range of molecules could be labeled to be studied. The mere modification of some amino acids within the GFP chromophore has led to the production of an infinite number of variants spanning from the blue to the red spectrum. Currently, the range of applications of fluorescent proteins has crossed the frontiers of life sciences, being used in developing detectors of different ions and even explosives

The milestone of observation through the lighting given by GFP inside biological systems could be compared with the advent of the supply of artificial lighting in public spaces

Interestingly, in his adolescence, Osamu Shimomura was walking home from work when he was exposed to radioactive fallout from the atomic bombing of Hiroshima, located just over 10 km away from him. It is believed that his grandmother told him to take a shower when he arrived home, which may have proved essential to avoid suffering radiation-related illnesses. Around 63 years after that unjustified tragedy (as the scientist calls it) and 46 years after the discovery of GFP, Osamu Shimomura was awarded, along with Martin Chalfie and Roger Y. Tsien, with the Nobel Prize in Chemistry in 2008

Secondary fluorescence is often employed for the analysis of cellular and subcellular dynamics. It is especially useful in vivo and in vitro for the characterization of molecular localization, abundance, and distribution. The advent of green fluorescent protein (GFP) isolation from a jellyfish by Osamu Shimomura (Fig. 4.1) marked a highly relevant watershed in scientific research. Knowledge about this protein was employed to design hundreds of fluorescent labels in different colors that could be

4 The Widefield Microscope

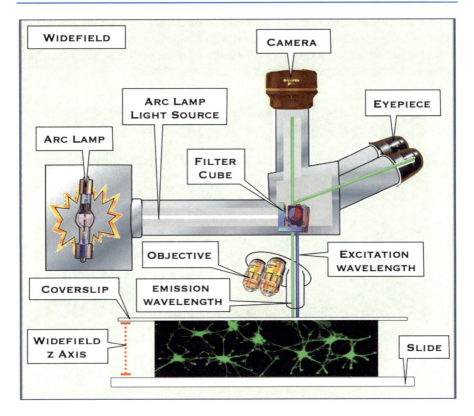

Fig. 4.2 Widefield, meaning
This type of microscope owes its name to how it obtains fluorescent information from the sample: via extensive areas along the Z-axis. Unfortunately, quantity is not quality; it is not possible to generate optimally focused optical planes when recovering so much light information and concentrating it in a single field of view; therefore, the sharpness of the image will be poor. In this microscope model, the energy coming from the arc lamp will radiate to the sample, so the fluorochromes will be excited and fluorophores will emit the corresponding fluorescence. The emission of light will occur in all areas of the sample where the lamp's energy can access. The light information from the sample that reaches the objective's front lens will then travel to the visual area of the eyepiece. Along the X- and Y-axis, the area that we will be able to observe will be limited by the magnification of the objective lenses and the eyepieces. The observed area will also include the Z-axis, corresponding to the thickness of the sample; this last characteristic generates a wide field of information, hence the naming

associated with many biological components within tissues and living cells such as proteins and nucleic acids, easily visible under fluorescence light microscopes which contain elements for selective excitation and to recover the fluorescence coming from these molecules.

The widefield microscope, a member of the fluorescence microscopy family, as was mentioned before, has been assigned that name due to the large amount of information it can collect along the Z-axis (depth or thickness of the sample or axial axis) (Fig. 4.2). This microscope's downside is its low efficiency when filtering a focal plane, which makes optimal focusing of specimens virtually impossible. To properly analyze optical planes of interest with a widefield microscope, it is necessary to adjust the micrometric screw. However, once the optical plane of interest is focused, the adjacent upper and lower planes will be blurred.

Dancing cells

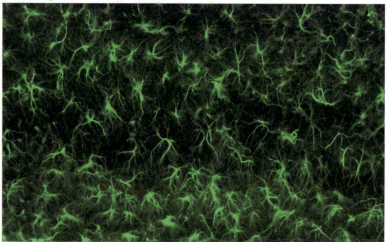

Astrocytes in the Dentate Gyrus, immunofluorescence for GFAP

The widefield microscope is also commonly known as a fluorescence microscope. This name is incorrectly used since fluorescence microscopes are a complete family within microscopy. A more accurate way to refer to it would be epifluorescence microscope. The prefix epi- is of Greek origin and means for something to be "on" or "above" (Fig. 4.3); thus, this name correctly refers to the spatial position of the lighting resource.

These microscopes were initially designed with a highly similar setup to that previously seen in transmitted light microscopy. In this system, high-energy light originating from an arc lamp (Fig. 4.4) is transmitted through the sample and goes alongside emitted fluorescence to the eyepieces, causing those beams to be directed straight into the observer's eye, which causes ocular injuries consequence due to a prolonged observation. Moreover, emissions of fluorophores are lower in energy compared to those of the arc lamp, and the crossing of light-transmitted information ultimately leads to confusing observations and conclusions. These conditions galvanized the creation of the current *epi* design.

4 The Widefield Microscope

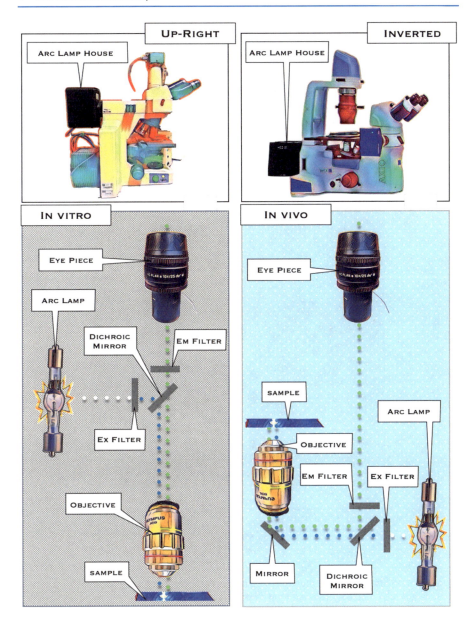

Fig. 4.3 Above fluorescent emission
This is what epifluorescence means; what is located here?—the arc lamp. In the conventional upright microscope configuration (left side of the figure), the lamp is located at the top where fluorescence emission occurs. When the microscope design is inverted (right side), which is useful for the observation of samples in vivo, the lamp will be located at the bottom. Note that the objective is also inverted; in this configuration, the specimen must be placed inverted as well, i.e., slides above and coverslips below. In both, upright and inverted configurations, once the light emission occurs, the light beam is directed to the excitation filter, then to the dichroic mirror. It then passes through the objective to reach the sample, where the fluorescence emission takes place. This emission is recovered by the objective and directed to the emission filter and finally to the eyepiece or a camera. Abbreviations: *Ex* excitation, *Em* emission

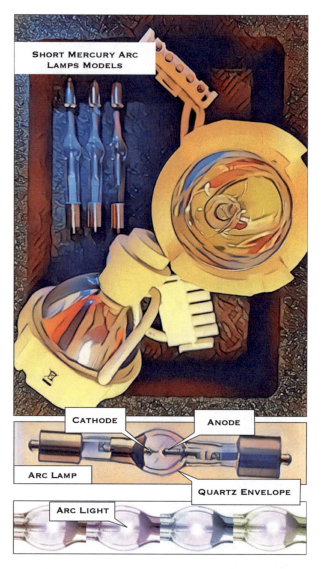

Fig. 4.4 Short arc lamps: a bright white-light resource
These lamps are commonly referred to as fluorescence lamps, a problematic name because they only emit high-energy white light, not fluorescent. An electric arc, also called a voltaic arc, is generated by an electrical discharge that originates from between two tungsten electrodes that are slightly separated from each other (hence the name short arc). These must be subjected to a potential difference and placed in an airtight atmosphere, which is given by a quartz glass bulb containing gas at high pressure (approximately 300 atmospheres or more). The illumination generated by these lamps can be 10–100 times more intense than that of an incandescent lamp (which is used in brightfield microscopes). The gases commonly used in these lamps are mercury and xenon, or a combination of both. Although arc lamps contain only a small amount of mercury or xenon, it is still important to avoid breakage of the lamp in order to prevent environmental or health damage. This type of lighting resource can heat up to reach between 200 and 300 °C. Thus, the element

4 The Widefield Microscope

Fluorescent cell planting field

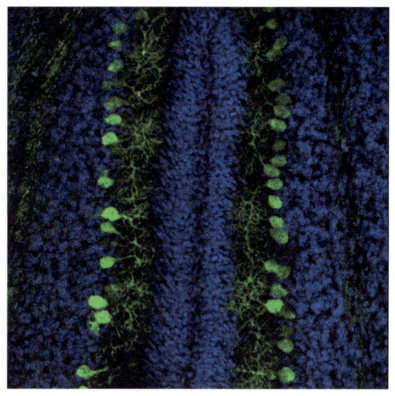

Purkinje cells in the cerebellum, immunostain for calbindin and DAPI stain

Obtention of the selective information provided by the various fluorescent labels in widefield microscopy is supported by three strategic elements located within the microscope: (1) excitation filter (or polarizing excitation filter), (2) dichroic mirror filter (the word dichroic comes from the Greek dikhroos and means "two colors"), and (3) emission filter (or polarizing emission filter) (Fig. 4.5). This system composed of two polarizing filters, and a mirror-like filter continues to be improved throughout the years in order to further limit and specify the employed excitation ranges, as well as to improve the precision when recovering emission wavelengths.

Fig. 4.4 (continued) within will transiently transform into plasma. Once the plasma is stabilized, the illumination is usually uniform. To favor optimal lightning, it is recommended to turn on the arc lamp at least 20 min before using the microscope and not turn it back on again within 30 min of the last use. Due to the high temperatures emitted by the arc lamp, the positive (anode) and negative (cathode) poles of the lamp deteriorate, as well as the elements of the microscope that are in contact with the high energy. This is why there are strategies to increase lens durability, such as the application of the photolithography mentioned briefly at the end of Chap. 1

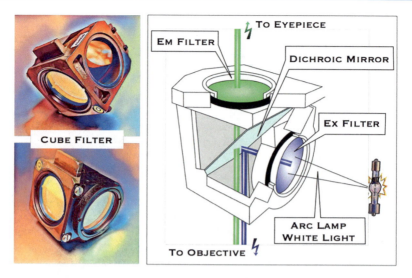

Fig. 4.5 The heart and soul of the widefield microscope
Known as fluorescent filter cubes, these are responsible for filtering the light coming from the arc lamp. As this is white light, it contains a wide range of wavelengths, ranging from the infrared to the ultraviolet as well as the whole visible spectrum. All wavelengths will be initially directed together to the excitation filter (Ex), and this element is designed to separate a specific wavelength from the many that make up white light. Then, this specific wavelength will be directed to a dichroic mirror, which is designed to reflect it and direct it to the sample for subsequent production of fluorescence. The fluorescent light emitted by the fluorophores will have a different wavelength than the excitation light. This emitted light is captured by the objective front lens and returned to the dichroic mirror, which now functions as a polarizing filter allowing the passage of the wavelengths that correspond to the chromophore. Finally, the filtered wavelength will be directed to the emission polarizing filter (Em), where many wavelengths that do not correspond to the desired emission spectrum will be obstructed before reaching the eyepieces

As in bright light microscopy, widefield microscopy recognizes that the interactions between the light and the sample itself result in various highly interesting phenomena (Fig. 4.6).

4 The Widefield Microscope

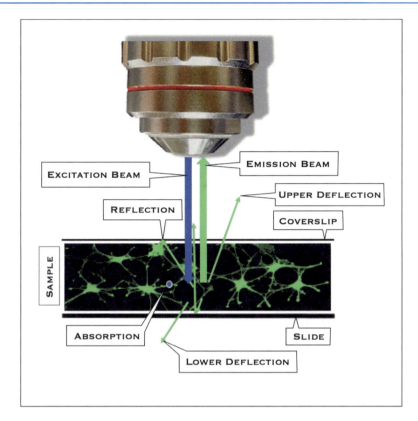

Fig. 4.6 From high-energy excitation to low-energy emission
In fluorescence microscopy, there are two events of relevance: excitation and emission. When a beam of a given wavelength (correspondent to the emission filter) collides with the sample, the subsequent radiation will excite fluorophores to its same wavelength, which will result in the emission of fluorescent light. The energy corresponding to light excitation is always greater than that corresponding to the subsequent emission. At any given point during excitation and emission, there can occur reflections, absorptions, and deflections of light from the primary beam. Also, fluorescence emission may include primary emission signals (autofluorescence)

Green fluorescent flowers reflected in the water in a night landscape

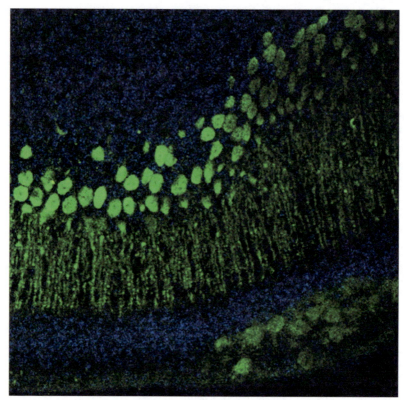

Pyramidal cells in the hippocampus, immunofluorescence for tubulin and DAPI stain

The white-light beams emitted by both arc and tungsten lamps are within the visible spectrum, typically 405–691 nm (as seen in Chap. 2); the difference between the two is that the light from an arc lamp will be of greater energy and luminosity. A widely known fact is that when the lighting resource is more energetic, the resolution will consequently improve. Thus, widefield microscopes resolve 250 nm (providing optimal conditions for both the equipment and the sample) thanks to the use of an arc lamp. This is almost double the resolution reached by white-light microscopy through the use of tungsten, halogen, or LED resources, which sit around 400 nm.

In addition to the previously mentioned mercury arc lamps, xenon arc lamps exist as an alternative for widefield microscopes. The former are the most widely used due to their lower energy consumption (which consequently makes overheating highly uncommon) and prices. However, they generate variable maximum and minimum peaks for different wavelengths, so the lightning beams are less uniform when compared to those of xenon lamps, which provide more stable fluxes along the visible spectrum and are quick to stabilize after being lit. Nevertheless, their higher pricing and lower lifespans are disadvantages to consider.

As it might be obvious to you already, the entity responsible for fluorescence emission and the wide range of tones evoked within the visible spectrum is the fluorochrome–fluorophore–chromophore complex. Figure 4.7 shows fluorochrome's cycles of excitation and emission (also known as on or off cycles), and an overall explanation of how fluorescence is attained.

The amount of evoked excitation–emission cycles is inherent to the employed fluorochrome. Some have an inherent lower photostability, while others have been designed to be more stable and can stand a large amount of cycles. Regardless, two phenomena that are related to the intensity of energy received, as well as the time of exposure to the arc lamp, will inevitably occur in the fluorochrome. The first one is quenching, and it refers to the reduction of the excitation period or the lifetime of a fluorochrome. It does not translate into the total extinction of the emitted fluorescence, but rather into a decrease of intensity. The second one is photobleaching (or fading), which consists of an irreversible modification of the fluorochrome's molecular structure and implicates a full extinction of fluorescence. Temporally speaking, quenching will occur before photobleaching does.

Green varicosity

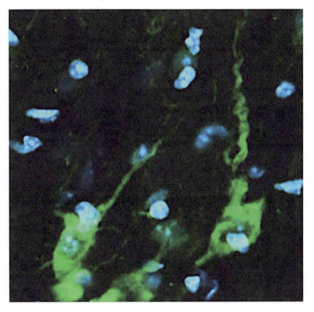

Interneurons of the neostriatum, immunostain for ChAT

It is thanks to fluorescent proteins, also called chimeric proteins or fusion proteins, that we have finally stopped being blind to cellular and subcellular events; we have finally stopped observing cellular physiology indirectly! Let us remember that when using a white light microscope (a predecessor of the widefield microscope), to carry out enzymatic colorimetric staining the specimen had to be dead. The advent of improvements in super-resolution fluorescence microscopy allowed for the occurrence of phenomena that revolutionized the field of microscopy; we will hear about them in the near future.

For widefield microscopes, the objectives are specially designed to take advantage of the high-energy emissions of mercury or xenon arc lamps. An objective can

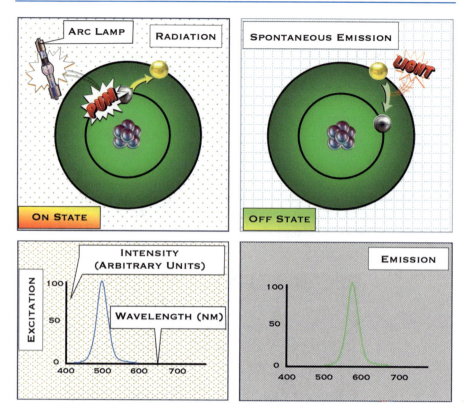

Fig. 4.7 Excitation and emission in detail

Up: Fluorescence occurs when a fluorochrome absorbs high-energy radiation emitted by a source, in this case, an arc lamp. Subsequently, the electrons of the fluorophore are excited by the received radiation, which means they will now contain a different amount of energy than their initial basal one. This will occur in a state known as "on-state." At this time, at a subatomic level, the energized electrons change orbital. To return to their basal or ground state, which is necessary to preserve the stability of the fluorochrome, they must dissipate the energy gained and then reposition in their original orbital. The dissipated or released energy will manifest itself as spontaneous light emission (fluorescence). This light (given by the fluorophore) has the characteristic of corresponding to a lower energy than the excitation light. This last phenomenon of energy release has been called the "off-state." Therefore, on-state corresponds to the excitation of the fluorochromes and is generated by the irradiation of light by the arc lamp. The excitation wavelength is selected with the support of an excitation polarizing filter. On the other hand, the off-state corresponds to the emission of light given by the fluorophore. Emitted light within a specific spectrum is made possible by the chromophore and separated thanks to the emission filter. Both the on and off states are quantified in nanometers, as these are the correct units for wavelengths. Below: As each fluorochrome designed today has a specific excitation and emission wavelength, it is important to make note of them before observing any sample under a microscope of the fluorescence family. This diagram is based on Stokes' observation

recover primary as well as secondary fluorescence emissions from the samples. For instance, Neofluar objectives allow for optimized transmission of wavelengths close to the range of ultraviolet light. Their design can reduce the transmittance of low-energy signals associated with autofluorescence and recuperate large amounts of light, making them ideal for widefield microscopy.

Finally, some of the advantages of the widefield microscope include the ability to generate images in vitro and in vivo, high contrast, and high specificity. Unfortunately, it does have disadvantages, including rapid photobleaching, common light dispersion, autofluorescence, and production of low-resolution 2D images. This last disadvantage is glaringly evident when compared to any other microscope in the fluorescence family.

To integrate the most notable components of an epifluorescence microscope, go back to Fig. 4.3, your look will surely be different than when you analyzed it for the first time.

4.1 When Should a Widefield Microscope Be Used?

A widefield microscope is used to obtain two-dimensional micrographs of specimens containing molecules that emit fluorescence at specific wavelengths (either *in vivo* or *in vitro*). It is an appropriate choice for determining the presence or absence of a molecule labeled with a fluorochrome at the cellular or subcellular level. Examples include immunofluorescence techniques, in situ hybridization, and even the observation of cellular dynamics such as calcium signaling.

Although it has limited resolution, its ability to visualize certain organelles or other cellular components is greater than that of a brightfield microscope. Like the latter, the optimal sample thickness ranges between 20 and 30 micrometers.

Widefield micrographs are suitable for fluorescence intensity analysis and for the extraction of other quantitative or semiquantitative data, including the localization and distribution of fluorochrome-fused molecules. However, widefield microscopy is generally not recommended for imaging that requires resolution below 300 nm, detailed descriptions of cellular structures, three-dimensional reconstructions, or certain fluorescence emission analyses (such as colocalization, total internal reflection fluorescence, and Forster resonance energy transfer) that require precise acquisition of emission events in a single optical plane. This concept will be detailed in the next chapter.

Further Reading

For GFP: Aliye et al. (2015), Ferreira et al. (2022), Prasher et al. (1992), Shimomura (2005), Weinstein et al. (2023), Weiss (2008), Zhang (2009), and Zimmer (2009).

For fluorescence microscopy (widefield microscopy) basis: Aswani et al. (2012), Disapro (2010), Pawley (2006), Sanderson et al. (2014), Stemmer et al. (2008), Verdaasdonk et al. (2014), and ZEISS (2023).

For fluorophorochromes, fluorophores and chromophores: Meaney and McGuffin (2008) and Merck (n.d.).

For fluorescence and autofluorescence (secondary fluorescence): Waldeck et al. (2009).

References

Aliye N, Fabbretti A, Lupidi G, Tsekoa T, Spurio R (2015) Engineering color variants of green fluorescent protein (GFP) for thermostability, pH-sensitivity, and improved folding kinetics. Appl Microbiol Biotechnol 99(3):1205–1216. https://doi.org/10.1007/s00253-014-5975-1

Aswani K, Jinadasca T, Brown CM (2012) Fluorescence microscopy light sources. Microscopy Today 20:22–28

Disapro A (2010) Optical fluorescence microscopy: from the spectral to the nano dimension. ebook

Ferreira JRM, Esteves CIC, Marques MMB, Guieu S (2022) Locking the GFP fluorophore to enhance its emission intensity. Molecules 28(1):234. https://doi.org/10.3390/molecules28010234

Meaney MS, McGuffin VL (2008) Investigation of common fluorophores for the detection of nitrated explosives by fluorescence quenching. Anal Chim Acta 610(1):57–67. https://doi.org/10.1016/j.aca.2008.01.016

Merck (n.d.) Fluorophore vs Fluorochrome, in: flow cytometry dye selection tips. https://www.sigmaaldrich.com/MX/es/technical-documents/technical-article/protein-biology/flow-cytometry/how-to-select-flow-cytometry-dyes?srsltid=AfmBOorakwMvbP62ePBp6IbsJj8iBoVzh4vmf08RstHKBqk9zHNU8CEA. Accessed 22 Oct 2022

Pawley JP (2006) Handbook of biological confocal microscopy, 3rd edn. Springer Verlag, p 931

Prasher DC, Eckenrode VK, Ward WW, Prendergast FG, Cormier MJ (1992) Primary structure of the Aequorea victoria green-fluorescent protein. Gene 111:229–233

Sanderson MJ, Smith I, Parker I, Bootman MD (2014) Fluorescence microscopy. Cold Spring Harb Protoc 2014:pdb.top071795. https://doi.org/10.1101/pdb.top071795

Shimomura O (2005) The discovery of aequorin and green fluorescent protein. J Microsc 217(Pt 1):1–15. https://doi.org/10.1111/j.0022-2720.2005.01441.x

Stemmer A, Beck M, Fiolka R (2008) Widefield fluorescence microscopy with extended resolution. Histochem Cell Biol 130(5):807–817. https://doi.org/10.1007/s00418-008-0506-8

Verdaasdonk JS, Stephens AD, Haase J, Bloom K (2014) Bending the rules: widefield microscopy and the Abbe limit of resolution. J Cell Physiol 229(2):132–138. https://doi.org/10.1002/jcp.24439

Waldeck W, Mueller G, Wiessler M, Brom M, Tóth K, Braun K (2009) Autofluorescent proteins as photosensitizer in eukaryontes. Int J Med Sci 6(6):365–373. https://doi.org/10.7150/ijms.6.365

Weinstein JY, Martí-Gómez C, Lipsh-Sokolik R et al (2023) Designed active-site library reveals thousands of functional GFP variants. Nat Commun 14:2890. https://doi.org/10.1038/s41467-023-38099-z

Weiss PS (2008) 2008 nobel prize in chemistry: green fluorescent protein, its variants and implications. ACS Nano 2(10):1977. https://doi.org/10.1021/nn800671h

ZEISS (2023) ZEISS campus fundamental concepts underpinning fluorescence microscopy. https://zeiss-campus.magnet.fsu.edu/articles/basics/fluorescence.html. Accessed on 14 Feb 2023

Zhang J (2009) The colorful journey of green fluorescent protein. ACS Chem Biol 4(2):85–88. https://doi.org/10.1021/cb900027r

Zimmer M (2009) GFP: from jellyfish to the Nobel prize and beyond. Chem Soc Rev 38(10):2823–2832. https://doi.org/10.1039/b904023d

The Confocal Microscope

5

There are many microscopes, but few microscopists.

Glenn Richards

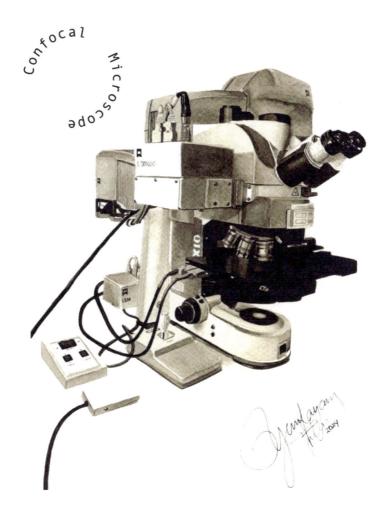

© The Author(s), under exclusive license to Springer Nature Switzerland AG 2025
A. Rosas-Arellano et al., *Microscopic Wonders*,
https://doi.org/10.1007/978-3-031-92559-7_5

The metamorphosis: Cells in cocoon

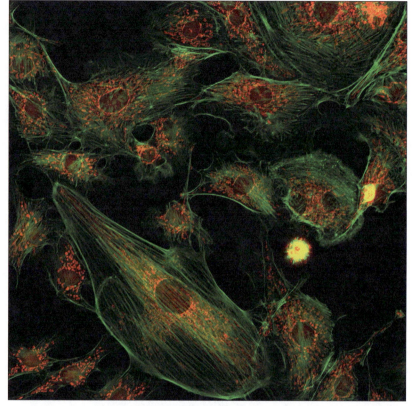

Lung cells, Immunostaining for tubulin and HSP60

Commonly known as a confocal microscope or confocal laser scanning microscope, this instrument is designed to increase resolution and optimize the capture of high-contrast, sharp, and highly detailed fluorescence images. Thanks to its innovative technology the family of fluorescent microscopes has been consolidated as indispensable for the observation and precise description of phenomena at the cellular and subcellular level. The resolution difference achieved by a confocal compared to a widefield microscope is around 50 nm, the former being capable of reaching a resolution limit of ~200 nm.

The elements that distinguish the confocal microscope are diverse, among which four main ones stand out: (A) high-intensity laser light as the illumination resource, (B) galvanometer mirrors, (C) pinhole, and (D) photomultiplier.

A. *Laser light*: The word "laser" is an acronym of five words and means *light amplification by stimulated emission of radiation.* How does this light amplification occur? Figure 5.1 explains this phenomenon and compares it with the spontaneous emission given by widefield microscopy using simple diagrams.

5 The Confocal Microscope

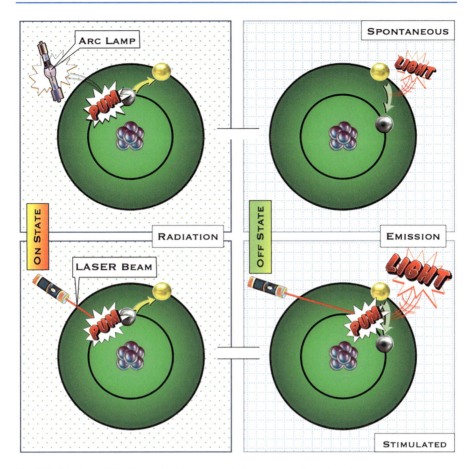

Fig. 5.1 Light amplification just at the moment of emission using a laser beam
The emission of fluorescence using a laser is similar to that produced by an arc lamp, that is, a photon of laser light that radiates an electron with an energy intensity equal to the difference between the basal orbital of the electron and the next higher orbital. This will position the irradiated electron at a higher energy level in which it will be highly unstable: the excitation or on-state. In a matter of micro- or nanoseconds, the electron will return to the more stable basal state by releasing spontaneously the excess energy as fluorescence (off-state). The difference between the emitted fluorescence that stems from the use of an arc lamp (upper panels) rather than from a laser (bottom panels) is that in the latter, there is an amplification of the emitted light. This is known as light amplification through emission stimulation, and it occurs when a photon from the laser beam irradiates an electron right at the moment of fluorescence emission, which results in greater amounts of light (light amplification). This way of describing is calling George G. Stokes' observations

> Laser light is within the spectrum visible to the human eye; this spectrum corresponds to white light and the wavelengths that comprise it and that are perceptible to us. The ranges covered by the spectrum go from ultraviolet to infrared, and between them are the wavelengths corresponding to blue, cyan, green,

yellow, and orange (Fig. 5.2). Are these colors the only ones we can perceive? It is considered that the healthy human eye is capable of distinguishing more than a million color shades. The colors that do not appear to be found in the spectrum and that are also perceptible to the human eye are explained by the transition quality of wavelengths and therefore by the combination of the colors contained within visible light. For instance, amber is generated by the combination of two colors, orange and yellow.

If laser light is visible as well as light coming from any other white light resource, what are the characteristics that distinguish them? A laser light beam has the following characteristics: it is monochromatic and highly energetic and has high directionality and coherence, which is something that could be loosely simplified as the movement of the laser light beam through the electromagnetic spectrum in a highly organized manner (Fig. 5.3).

Inspired by the experiment of Carbone and his team (Chap. 2, Fig. 2.4), we took it upon ourselves to photograph with a confocal microscope the green light of a laser. We believe this is the very first time a photo showing the high directionality and organization of laser light waves has been taken (Fig. 5.4).

Fig. 5.2 (continued) The visible spectrum corresponds to the wavelengths we can observe. They range from ultraviolet to infrared, and they go from highest to lowest energy, correspondingly. All of the different tones correspond to the decomposition or dispersion of white light and range in the average human eye from 360 to 780 nm (upper panel). The green tones located in the center of the spectrum are the easiest for the human eye to capture, making them "friendly" for our eye. It is therefore not a surprise that visiting places with a great abundance of green tones make us feel relaxed, such as natural areas and gardens. Is it like that for you when you observe these photographs? Which of them would produce the greatest state of relaxation when imagining yourself walking on this path? Do you agree that nobody says—I look for an ultraviolet-infrared forest, just to feel relaxed

5 The Confocal Microscope

Fig. 5.2 The colors seen by the human eye

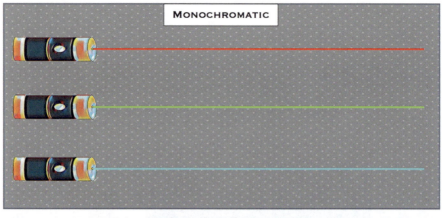
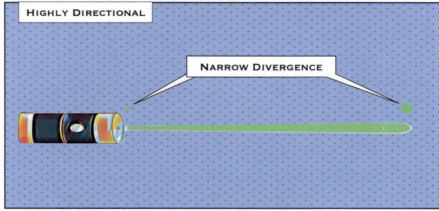
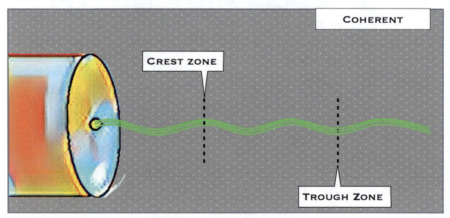

Fig. 5.3 Organized, high-energy light

Given its monochromatic nature, a laser beam generally emits light of a single color, although a few years ago a white laser was developed that could be broken down into different wavelengths. Additionally, laser light shows high directionality, which is why laser pointers are useful teaching aids for presentations. It is a highly energetic light, which is why its use for recreational accessories has been restricted globally for safety concerns. Finally, it shows high coherence, which makes it a highly ordered light. This characteristic implies that the crest and trough of the light waves that compose it are coordinated with each other, generating crest–crest and trough–trough arrangements

5 The Confocal Microscope   77

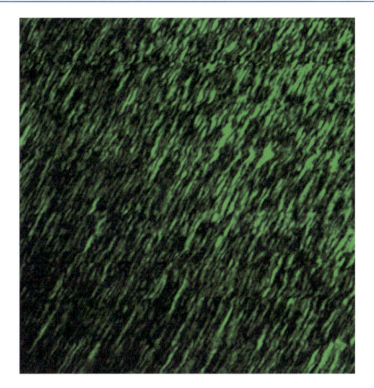

Fig. 5.4 Laser waves micrographed with a confocal microscope
A laser light with a wavelength of 532–650 nm and 100 mw was projected onto a slide containing distilled and deionized water and photographed using a ZEISS LSM710 confocal microscope. The capture conditions of the micrograph were the following: water immersion objective 20×/1.0, laser 488 70%, pinhole 1 AU, master gain 570 V, offset −5425, and digital gain of 2. Image obtention was supported by the ZEN black software, version 2012 SP4 (Carl ZEISS, Jena, Germany). As controls, micrographs of 561- and 647-nm laser lines were also attempted, as well as photographs with no laser or no water. In all these cases, no signal was seen (not shown), confirming that the waves in the left photomicrograph correspond to the light cast by a green laser

The Van Goh's cells

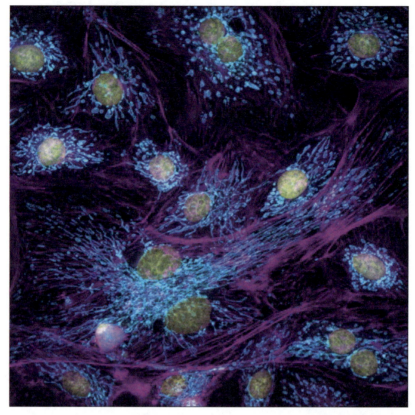

Lung cells, Immunostaining for tubulin and HSP60, DAPI

B. *Galvanometer mirrors*: These elements direct the laser beams in a highly precise manner to perform a sample reading spot by spot. If the image is required to be of higher resolution, it will be necessary to use a longer capture time for each pixel, which is known as "pixel dwell time," and/or increase the number of pixels assigned for image formation, which would increase precision. There are usually two galvanometer mirrors, and they have the function of deflecting the laser beam at a certain angle to make it hit the sample very finely in a reading procedure called "scanning." One of the mirrors takes a reading of the horizontal plane while the other does so in the vertical plane. This is why this microscope is known as a confocal laser scanning microscope.

Although galvanometer mirrors have a high rotation speed, the reading time of a confocal microscope high-quality images will also depend on factors such as (1) laser scanning speed (scan speed), (2) image size, for example, 1024 lines per 1024 lines of reading (means 1024 × 1024 square pixels per frame), and (3) programmed repeated readings per pixel (2×, 4×, 8×, or 16×) (Fig. 5.5). However, when looking to improve image quality, it must also be considered that slow readings, high

5 The Confocal Microscope

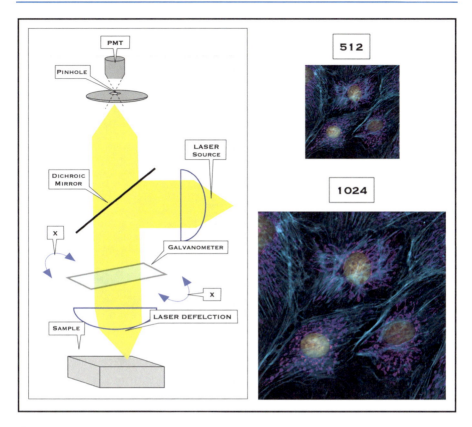

Fig. 5.5 High-speed and precision technology
Galvanometer mirrors work in x–y coordinates, and their function is to perform a point-by-point laser scan of the sample until a complete image frame is completed (left). Due to the high speed and precision, a micrograph with high contrast and resolution is obtained. The most common formats that make up a confocal photomicrograph are 512 × 512 pixels and 1024 × 1024 pixels, which equates to 262,144, and 1,048,576 reading points, respectively (right). Abbreviation: PMT, photomultiplier. PMT and pinhole are explained later

pixelation, and use of repeated readings could result in long periods of capture time, and if the photostability of the employed fluorochromes is not considered, fluorescence emission could be affected.

C. *Pinhole*: This functional accessory could be considered one of the most important physical elements within the confocal microscope, and what differentiates it from the rest.

Before proceeding to describe the pinhole, it is essential to understand the foundation of this element, which is the "Point Spread Function." This function was first described by George Biddell Airy. It elegantly explains that if a beam of light is projected onto a physical barrier with a tiny pinhole made in it, we will see a round

Fig. 5.6 The disk of Airy: an additional message about resolution limits
A very small central spot of light surrounded by a diffraction pattern was described by Sir George Biddell Airy in the 1820s. This is a representation of an Airy disk applying a negative style to highlight the significant density of the central disk and the low density of its concentric rings as described by Airy. Always, the disk diameter is closely related to two elements: (a) light wavelength and (b) hole opening size. The description of this tiny hole and the light that passes through it gives an extra message: resolution is not only related to light wavelength, the media, and lenses but also related to the sample and its interaction with the central axis of the light; in this area, a better resolution is obtained compared to the diffracted light regions (light scattering rings or areas that do not correspond to the axial axis of the light)

light projection onto the screen placed on the other side. Interestingly, we will see a concentric of lower light density surrounding this circle. If we are more observant, we will notice that this ring is accompanied by others around it and that they will always be of lower density than the one preceding it.

This phenomenon has already been described in Chap. 2 and corresponds to the wave behavior of light and its dispersion. So, unlike the double slit described by T. Young, what G. B. Airy was describing was the scattering of light through a tiny round-shaped hole. This tiny projected circular beam of light has been named Airy's disk (Fig. 5.6).

Likewise, the units with which the pinhole aperture diameter is adjusted in a confocal microscope have been named Airy units. Appropriately using this knowledge when employing confocal microscopy is key to achieving images with optimal resolution, making it easier to avoid misinterpretations of information.

The knowledge inherited by G. B. Airy is used in the machinery of the modern confocal microscope. The fluorescence emitted by all sample areas reached by the laser beam will be captured by the front lens objective and will be redirected within the microscope through various lenses and mirrors, finally reaching a camera specialized in capturing fluorescent light information. Up to this point, there seems to be no difference between how widefield microscopes and confocal microscopes acquire light information. However, before the light reaches the capture system in a confocal microscope, there is a tiny diaphragm: the pinhole. This small mechanical system has the function of blocking out-of-focus scattering of light, limiting the information recovered to narrower focal planes, and allowing for better axial resolution (Fig. 5.7).

Being a dynamic system, the pinhole works with various apertures, and the length of the aperture is measured in Airy units, as mentioned above. When the pinhole is completely open, our confocal microscope is essentially a widefield microscope coupled to a laser light source. This is because we are allowing the passage of fluorescent light coming from all Z (sample thickness) plane axis. Seen in another way, the fluorescent information that we are recovering with the pinhole fully open is coming from all, in-focus, over-focus, and under-focus optical planes. Closing the pinhole aperture allows us to control the amount of information integrating the photomicrography by limiting the light coming from the Z-axis, without affecting information coming from X and Y.

The resolution of the information coming from the X- and Y-axes is known as the lateral resolution. On the other hand, the axial resolution is given by light information recovered from the Z-axis of the sample. The latter will always have a lower power than the first. This is because it requires the laser to penetrate the sample and for light emissions to pass through its thickness before being recovered by the microscope without diffraction, a situation that does not occur. On the contrary, lateral resolution encompasses only surface information, that involves lesser light scattering.

So far, we have mentioned why opening the pinhole beyond its optimal aperture directly affects the resolution of the image and makes the confocal microscope work like a widefield microscope. But what happens if we close this diaphragm beyond the optimal aperture? Figure 5.8 illustrates this scenario. Since the elimination of out-of-focus planes is one of the main advantages of the confocal microscope, optimal narrowing of the pinhole allows for the observation of molecule colocalization (provided they are labeled with different chromophores) with a high degree of accuracy. Colocalization studies look at molecules located close together in the same space at an instant in time. This is useful because physically connected molecules often co-participate in the same biological mechanisms.

Furious fire

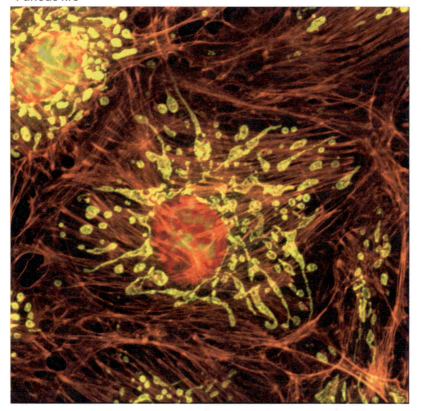

Lung cells, Immunostaining for tubulin and HSP60, DAPI

In a colocalization study, the magnification employed is determined based on the size of the molecules of interest. Objectives with higher magnifications result in thinner optical sections, which improves resolution and increases accuracy when analyzing colocalization information. In Fig. 5.9, you can appreciate the differences between the optical cuts of the Z dimensions using a 63× objective against a 20× objective (excitation wavelength of 488 nm). Even when using an objective with high magnification, it is necessary to monitor the pinhole aperture, which is optimal at one Airy unit for colocalization studies. Beyond this value the optical sections with become larger, increasing the probability of capturing colocalization signals, but decreasing accuracy.

Finally, it is important that in a colocalization study fluorochromes used have excitation spectra that do not overlap with each other (as mentioned at the end of this chapter) and thus cannot be accidentally coexcited and provoke false positives.

The capacity of the confocal microscope to select specific optical planes for capture allows us to group serial optical sections captured separately to form three-dimensional models that can provide useful information when looking at length,

5 The Confocal Microscope

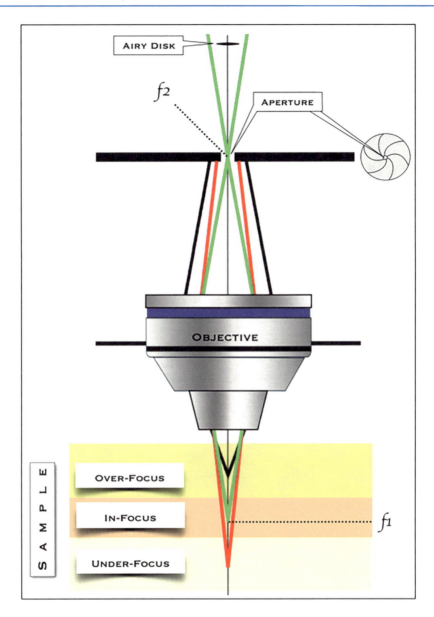

Fig. 5.7 Conjugate focal plane microscope
Although not employed in the English language, this could be another way to refer to the confocal microscope. What would "focal conjugate" refer to? In this diagram, you can see the sample area divided into three focal sectors: under-focus (light yellow), in-focus (orange), and over-focus (dark yellow). Fluorescence emission can occur in all these areas and is represented in red for under-focus, green for in-focus, and black for over-focus. The objective front lens can be seen above the sample. It is here where the information will be captured and redirected toward the pinhole. The pinhole opening will allow the passage of the in-focus information but will "eliminate" the areas that are not in a conjugate plane (the black and red lines). Therefore, the pinhole aperture is considered a focal plane (f2), and it is conjugated with the other coming from the light emission of the sample (f1). Confocal is what we refer to this mvicroscope as a short name

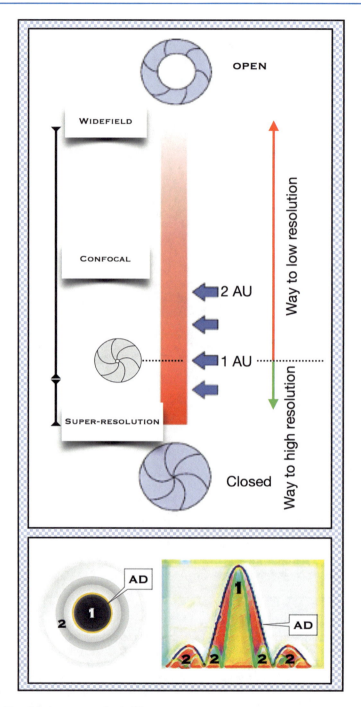

Fig. 5.8 The pinhole power under 1 AU

5 The Confocal Microscope

depth, location, and distribution. This is how the 3D function, another of the main characteristics of this type of microscope, arises (Fig. 5.10).

The existence of this function leads us to question how the thickness of an optical slice is determined and how the confocal system establishes the distance between slices in the Z-axis to form reliable 3D images. This is done using data from the pinhole aperture, the excitation wavelength, the immersion medium (such as air, water, and oil), and the numerical aperture of the objective. This is one of the reasons why you should never forget to adjust the pinhole to the optimal every time you change fluorochrome channel or objectives; because this diaphragm is dependent on the numerical aperture and light wavelength as well, the former is determined by the objective in use. The manual or automatic registration of the lens in the microscope software will also provide related information about the immersion medium, while the wavelength would have been recorded since the beginning of the capture parameter assignment.

Therefore, the dimensions at the Z planes of the thickness of the captured optical plane and the distance to start a new capture are always dependent on the elements that have been mentioned. A useful advice would be to consider that if the analysis of molecules requires specific dimensions, the combination between the objective and employed wavelength will have to be analyzed to obtain greater confidence when capturing and reporting a specific biological phenomenon. Blue and green wavelengths provide greater resolution than red ones.

Although G. B. Airy determined the Point Spread Function on a tiny circular hole, we can currently find grid-shaped pinholes in some models of confocal microscopes (Fig. 5.11).

When you want to reconstruct a structure three-dimensionally, it is desirable to use a 1-AU pinhole besides that the confocal software employed can assign the optimal optical planes and the spacing between them to carry out an optimal reconstruction, otherwise, oversampling or subsampling could occur in manual mode, altering the final interpretation of an experiment, generating data that will not be reproducible (Fig. 5.12).

Fig. 5.8 (continued) Upper: The optimal pinhole opening is assigned as one unit of Airy (1 AU). The area or diameter of the aperture of 1 AU is dependent on the wavelength, magnification, and the numerical aperture of the objective used. An Airy unit ensures that the confocal system is optimized to acquire high-contrast and resolution images with a focal plane thickness optimized for the selected fluorochrome and objective. Opening the pinhole means there will be a greater amount of in-focus and out-of-focus information while reducing the aperture to less than 1 AU would suggest a loss of information. However, in cases where fluorescence is excessive, either due to high emission of the fluorophore or a wide distribution of the molecule containing the fluorochrome (as often occurs with highly expressed proteins) reducing the pinhole aperture to less than 1 AU favors resolution. In some cases, it can even exceed the 200-nm resolution achieved by a conventional confocal microscope, thus approaching the realms of super-resolution. Bottom: Improving resolution by decreasing the pinhole opening is effective because the Airy disk (AD) region is not homogeneous—it also contains areas of higher photon concentration (1), located right on the axial axis (larger yellow area inside of Gauss distribution), as well as lower photon concentration (2)

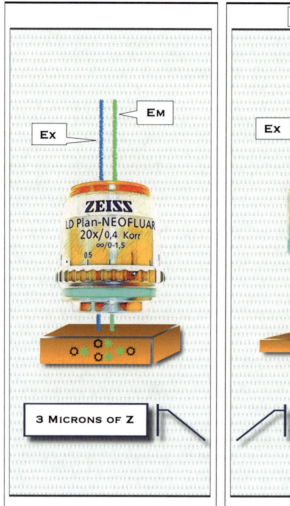

Fig. 5.9 A fit line is key
Conducting colocalization studies requires a dynamic imaging strategy. In the image, hypothetical laser beams with an excitation wavelength (Ex) of 488 nm reach the sample generating a fluorescence emission (Em). When an objective with a lower magnification (20×, left) is used, the thickness of the optical section will be greater, therefore the emission areas (cells, organelles, or molecules) will be wider, which favors erroneous colocalization signals. On the other hand, the 63× objective (right) generates information over a thin area, thus reducing the possibility of erroneous information, as fluorescence is limited in the Z plane of one section. The best strategy is to choose an appropriate fluorochrome and optimal pinhole aperture for the employed optical section to generate planes that are close to the dimensions of the molecules being studied

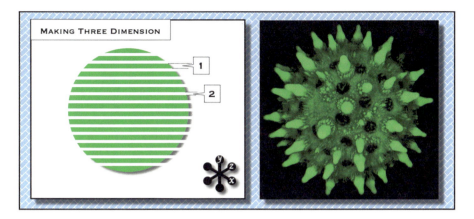

Fig. 5.10 Building 3D image
The 3D function involves the digital superimposition and reconstruction of serial optical sections. The X–Y plane, also known as 2D, is not truly two dimensional. If we look at it in detail, an optical plane implies a micrometric depth, that is, a Z plane. Therefore, we have two areas in Z that the software considers to create a 3D image: (1) The thickness of each X–Y optical plane. (2) The distance or space that exists between each of the X–Y planes that make up the final image. The sum of the thickness of a 2D plane, the optical jumps between each micrograph, and the number of 2D planes gives the total dimension in Z to the final 3D image (right). The final dimensions of a three-dimensional reconstruction can be represented in pixels or microns

Another of the prominent elements (besides (A) LASER, (B) galvanometer, and (C) pinhole) of the confocal microscope is the (D) photomultiplier, which is a highly specialized device capable of detecting light when it is scarce. It is described as a highly light-sensitive element capable of converting photons into measurable electrical signals for its subsequent amplification. In other words, any barely perceptible light in the environment can be measured and heightened to allow its observation. To achieve this, we will have to add an electrical charge to that barely perceptible light, a function that is generally performed by the photomultiplier within the confocal microscope. Photomultipliers are made up of a photocathode, a focusing system, a dynode chain, and an anode, among other elements (Fig. 5.13).

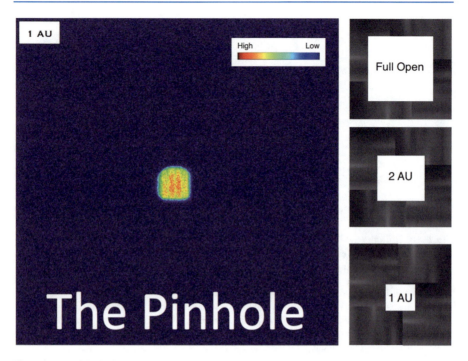

Fig. 5.11 A pinhole design: a square heart of a confocal from a well-known blue brand
Confocal microscopes display images with high contrast (high relation signal-to-noise ratio), improve resolution, and can produce three-dimensional images through optical sectioning (Z planes); these properties are the most prominent uses of confocal microscopy and are achieved thanks to the optimal application of the pinhole. Regardless of its round or square shape, this tiny diaphragm at 1 AU eliminates all out-of-focus light emission not originating in the focal plane allowing the recovery of highly ordered and concentrated photons (red dots in left); however, if the pinhole is open more than 1 AU (2 AU or completely open), optical thickness of the slice increases, signal-to-noise ratio decreases, and therefore the resolution, as well as the morphological quality of the 3D images, declines accordingly. Practical requirements such as in vivo imaging normally involve degree changes of confocality that require opening pinhole diameter; nonetheless, the full open aperture is considered a non-confocal image

Fig. 5.12 (continued) There are three general ways to create a 3D image. (1) *Optimal sampling*: The parameters are assigned by the microscope software and the dimensions of the reconstruction will be similar to reality. It is recommended to perform this function whenever detailed information is needed regarding the morphological characteristics of an organelle, cellular structure, cell, or tissue where the fluorescence of interest is located. (2) *Oversampling*: Given by a splicing of optical slices, its use is not recommended, since it tends to repeat information and deform structures. The parameters are regularly assigned manually by the experimentalist, and it is normally considered an erroneous capture. (3) *Subsampling*: A type of capture widely used when it is desired to know only the presence of a chromophore associated with a molecule; it does not imply the precise three-dimensional description of the structure where the fluorochrome is distributed or located since that would come with loss of information

5 The Confocal Microscope

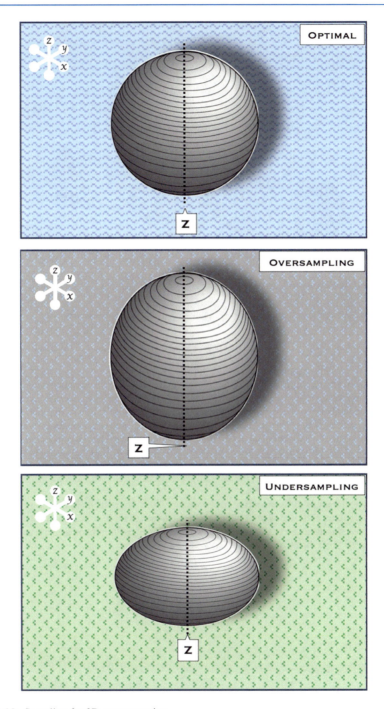

Fig. 5.12 Sampling for 3D reconstruction

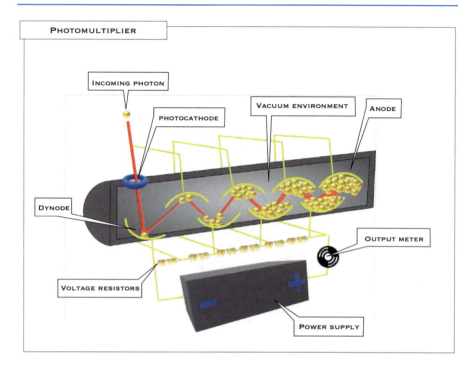

Fig. 5.13 From imperceptible to perceptible light

A conventional photomultiplier is made up of four main elements: (1) photocathode: transforms photons into photoelectrons (a negative charge is acquired); (2) focusing system: an electric field via which photoelectrons are accelerated and redirected; (3) amplification chain: made up of dynodes that contain a positive charge that increases progressively between dynodes and facilitates the attraction of more photoelectrons at each stage. The orientation of each dynode favors the chain mobilization of the photoelectrons produced in each of them, which in the example are produced arithmetically (2–4–8-16-32, etc.). The number of dynodes of each photomultiplier is model dependent; and (4) anode: captures the flow of photoelectrons and produces the output electrical signal measurable by the output meter

Currently, there are different types of photomultipliers. Some of them have been created to photomultiply in an outstanding way, generating avalanches of multiplied photons in a single step without the need to make an arithmetic chain, which favors the observation of the secondary fluorescence signal more than that of the primary one

5 The Confocal Microscope

Green splash

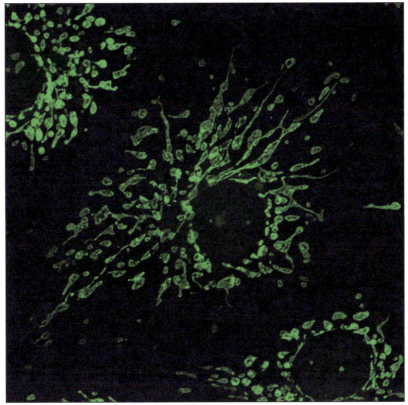

Immunostaining HSP60

In summary, some of the advantages of the confocal microscope are eliminating out-of-focus information, providing high contrast, high photomultiplication, and colocalization of information, as well as generation of 3D images. On the other hand, some of its disadvantages are rapid photobleaching, a relatively slow capture system (when it comes to high resolution or generation of 3D images in optimal quality), and that it is not suitable for high-speed *in vivo* captures. The functions of this microscope are broad, four of the specialized and popular techniques are briefly described in Figs. 5.14, 5.15, and 5.16.

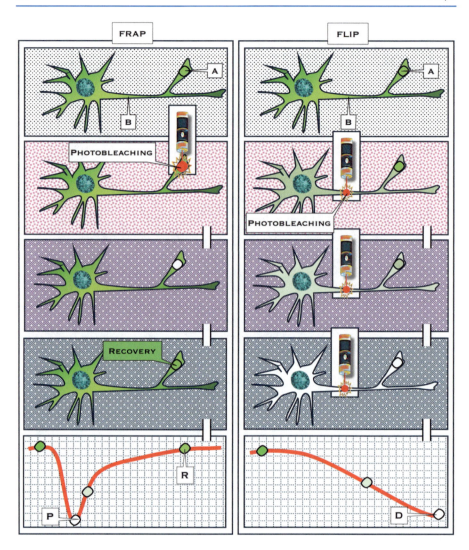

Fig. 5.14 Where do they come from and where are they going? FRAP and FLIP, two complementary microscopy techniques

Fluorescence recovery after photobleaching (FRAP): It is an in vivo recording technique that helps to expose in the lateral plane (x–y) the intracellular dynamics of molecules previously labeled with a fluorochrome. As shown in the example, if in a neuronal cell we want to know the expression site and the kinetics of a molecule that is located in a specific region or organelle (A) (first FRAP panel), a selective and short-term photobleaching of the region must be performed (second panel). If in a given time period of the FRAP recording, the photobleached area (third panel) gradually recovers fluorescence (fourth panel), there is a high probability that the synthesis and/or mobilization of the labeled molecule comes from a different area (B)

Fluorescence loss in photobleaching (FLIP): It is a technique to confirm whether a molecule that is supposed to be synthesized and mobilized from other cellular regions (B) comes from them or is synthesized in the study region (A); it is a technique that complements FRAP. The illustration shows that now (B) is the region that receives photobleaching (second FLIP panel), if during the

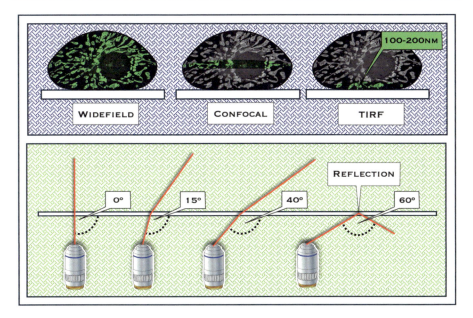

Fig. 5.15 TIRF: exploring the external side of the biological samples
In contrast to widefield microscopy, which obtains the emitted fluorescence from all areas where the excitation resource penetrates, and confocal microscopy, which obtains information from optical sections through the sample, total internal reflection fluorescence (TIRF) microscopy is a useful technique for recording cell kinetics coming from sample surface regions with a maximum depth of 100–200 nm (above). This microscopy is very useful when trying to know the size, abundance, distribution, localization, and kinetics of molecules labeled with a fluorochrome, specifically in membrane and submembrane regions

TIRF microscopy has been used to describe cellular vesicles and their dynamics (exocytosis and endocytosis), protein diffusion, neurotransmitter release, and the dynamics of the cell membrane and its components, among many other membrane-level processes. TIRF is commonly applied in inverted microscopes, although not exclusively, and is achieved thanks to an adaptation of the objectives (TIRF objectives) that usually have larger numerical apertures than standard objectives. TIRF objectives are designed internally with lenses adapted to direct the light source toward the sample at an angle of 60° (bottom) to produce total internal reflection (see Chap. 2). The higher numerical apertures of TIRF objectives allow capturing information from fluorescence emitted at a considerable angle of inclination; this information would be lost with a standard numerical aperture

Fig. 5.14 (continued) photobleaching of region (B) in zone (A) fluorescence quenching is also observed (as shown in the third and fourth panels), it is concluded that the labeled molecule is synthesized or diffuses from (B) to be located in (A). Unlike FRAP, which requires short-term photobleaching, in FLIP the photobleaching of region (B) is maintained throughout the recording. The graphs in the lower panels show the classical kinetics of FRAP and FLIP experiments. In the first, according to the illustrations in the upper panels, a decrease in fluorescence is observed until reaching photobleaching (P) and ending at the recovery time (R), while in FLIP the progressive decay (D) of fluorescence is represented, both representations are only for the study area (A)

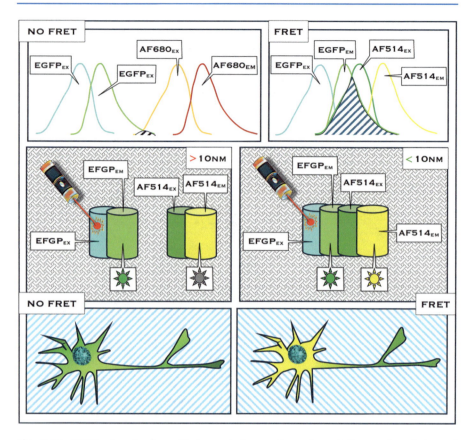

Fig. 5.16 FRET, the domino effect: a chain reaction of excitation and emission of fluorochromes

Forster resonance energy transfer (FRET) is a technique used to understand the interaction between two molecules labeled with different fluorochromes, but which have very close emission and excitation spectra. In FRET, the fluorochrome that is excited by radiation through a light source is called the donor and the other one that is excited exclusively by donor emission is called the acceptor. A FRET experiment can be performed on live or fixed samples. The top left panel shows the excitation (ex) and emission (em) spectra for enhanced green fluorescent protein (EGFP) and AF680 (Alexa Fluor 680). The black dashed region shows the junction between EGFP emission (donor) and AF680 excitation (acceptor). In terms of energy, it is insufficient to excite AF680; therefore, in this example, there is no FRET. The second panel (top right) shows the excitation and emission spectra of EGFP (donor) and AF514 (Alexa Fluor 514 [acceptor]); in this example, the binding area is considerable and therefore susceptible to FRET

There are two additional requirements for FRET to occur, in addition to the donor's emission spectrum interacting extensively with the acceptor's excitation spectrum. One is that both fluorochromes must be at a minimum distance from each other. A distance greater than 10 nm is not sufficient to activate the acceptor (left center), and distances less than or equal to 10 nm will successfully produce a FRET (right center). The other one condition is that the spatial position of the fluorochrome must be optimal to expose its active site to the energy emitted by the donor (not shown). The latter is a point to consider at the level of experimental results since even if the fluorochromes are as close to each other as 10 nm or less and their emission and excitation spectra are very close, if the conformation and position of the active sites of the fluorochromes are not adequate, FRET will not be performed. When a FRET is negative, only one emission is observed, and both when the FRET is positive, as shown in the final panels

The siamese cells

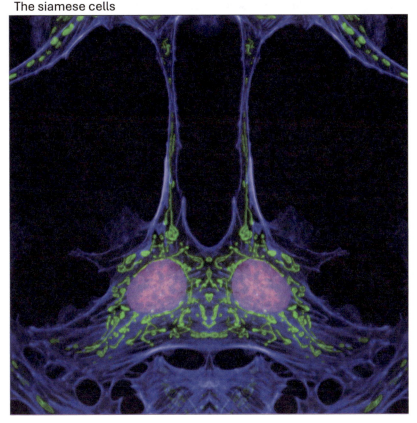

Immunostaining for tubulin and HSP60, DAPI

5.1 When Should a Single-Photon Confocal Microscope Be Used?

Its use is well-documented for spatial distribution and localization studies employing techniques such as immunofluorescence, FRET, FLIP, TIRF, calcium signaling, optical sectioning, colocalization (reliable at magnifications of 60, 63×, or more), and three-dimensional reconstructions in fixed tissue, live tissue, or at cellular specimens containing fluorescent molecules at specific wavelengths.

When equipped with modules such as *Airyscan* or *Spinning Disk*, the resolution typically reaches around 100 nm—twice that of a conventional confocal microscope. Although some models allow imaging of samples up to 500 micrometers thick, it is generally not recommended to use specimens of excessive thickness. This is because axial resolution is lower than lateral resolution, and it decreases with imaging depth. Even 100-micron sections can be challenging to reconstruct in 3D, making sample thicknesses between 20 and 60 microns ideal for this system.

Additionally, staining techniques like immunofluorescence have limitations due to the large size of some antibodies, which can hinder penetration. For conventional antibody sizes, a penetration depth of 8–9 microns has been reported in nervous system tissues. In some cases, this limitation can be overcome using nanobodies (nano-antibodies). However, preparing histological sections thicker than 20 microns with conventional antibodies is impractical unless adaptations to the immunofluorescence technique are implemented in order to enhance antibody penetration (some methods are described in Piña et al. 2022; see references).

Further Reading

For confocal microscopy basis: Bayguinov et al. (2018), Brzostowski and Sohn (2021), Hibbs (2004), Jerome and Price (2018), Mongan et al. (1999), Nwaneshiudu et al. (2012), Ockleford (1995), Paddock (1999), Smith (2001), and Wright and Wright (2002).

For TIRF microscopy: Fish (2009).

For colocalization: Jensen (2014).

For pinhole: Kitamura (2021) and Wilson (1995).

For objective lenses, numerical aperture and magnification: Piston (1998).

For nanobodies: Pleiner et al. (2018).

For visible spectrum: Sliney (2016).

References

Bayguinov PO, Oakley DM, Shih CC, Geanon DJ, Joens MS, Fitzpatrick JAJ (2018) Modern laser scanning confocal microscopy. Curr Protoc Cytom 85(1):e39. https://doi.org/10.1002/cpcy.39
Brzostowski J, Sohn H (2021) Confocal microscopy: methods and protocols. Springer. https://doi.org/10.1007/978-1-0716-1402-0
Fish KN (2009) Total internal reflection fluorescence (TIRF) microscopy. Curr Protoc Cytom 50(1):273–275. https://doi.org/10.1002/0471142956.cy1218s50
Hibbs AR (2004) Confocal microscopy for biologist. Springer. https://doi.org/10.1007/978-0-306-48565-7
Jensen E (2014) Technical review: colocalization of antibodies using confocal microscopy. Anat Rec (Hoboken) 297(2):183–187. https://doi.org/10.1002/ar.22835
Jerome WG, Price RL (2018) Basic confocal microscopy. Springer, Cham. https://doi.org/10.1007/978-3-319-97454-5
Kitamura A (2021) Pinhole closure improves spatial resolution in confocal scanning microscopy. In: Kim SB (ed) Live cell imaging. Methods in molecular biology, vol 2274. Humana, New York. https://doi.org/10.1007/978-1-0716-1258-3_31
Mongan LC, Gormally J, Hubbard AR, d'Lacey C, Ockleford CD (1999) Confocal microscopy. Theory and applications. Methods Mol Biol 114:51–74. https://doi.org/10.1385/1-59259-250-3:51

References

Nwaneshiudu A, Kuschal C, Sakamoto FH, Anderson RR, Schwarzenberger K, Young RC (2012) Introduction to confocal microscopy. J Invest Dermatol 132(12):1. https://doi.org/10.1038/jid.2012.429

Ockleford C (1995) The confocal laser scanning microscope (CLSM). J Pathol 176(1):1–2. https://doi.org/10.1002/path.1711760102

Paddock SW (1999) Confocal laser scanning microscopy. BioTechniques 27(5):992–996. https://doi.org/10.2144/99275ov01

Piña R, Santos-Díaz AI, Orta-Salazar E, Aguilar-Vazquez AR, Mantellero CA, Acosta-Galeana I, Estrada-Mondragon A, Prior-Gonzalez M, Martinez-Cruz JI, Rosas-Arellano A (2022) Ten approaches that improve immunostaining: a review of the latest advances for the optimization of immunofluorescence. Int J Mol Sci 23(3):1426

Piston DW (1998) Choosing objective lenses: the importance of numerical aperture and magnification in digital optical microscopy. Biol Bull 195(1):1–4. https://doi.org/10.2307/1542768

Pleiner T, Bates M, Görlich D (2018) A toolbox of anti-mouse and anti-rabbit IgG secondary nanobodies. J Cell Biol 217(3):1143–1154. https://doi.org/10.1083/jcb.201709115

Sliney D (2016) What is light? The visible spectrum and beyond. Eye 30:222–229. https://doi.org/10.1038/eye.2015.252

Smith CL (2001) Basic confocal microscopy. Curr Protoc Neurosci. https://doi.org/10.1002/0471142301.ns0202s00

Wilson T (1995) The role of the pinhole in confocal imaging system. In: Pawley JB (ed) Handbook of biological confocal microscopy. Springer, Boston. https://doi.org/10.1007/978-1-4757-5348-6_11

Wright SJ, Wright DJ (2002) Introduction to confocal microscopy. Methods Cell Biol 70:1–85. https://doi.org/10.1016/s0091-679x(02)70002-2

The Two-Photon Microscope

6

The physicist Maria Goeppert-Mayer laid the theoretical foundation for two-photon transitions in her 1930 PhD dissertation, but we will distill the essentials here. It turns out that it is possible to excite fluorescent molecules using two or more photons rather than a single photon.

Pina Colarusso and Craig Brideau

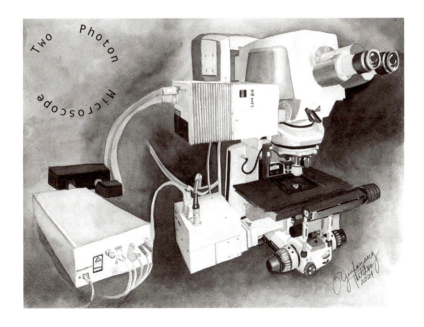

© The Author(s), under exclusive license to Springer Nature Switzerland AG 2025
A. Rosas-Arellano et al., *Microscopic Wonders*,
https://doi.org/10.1007/978-3-031-92559-7_6

Neurons work-meeting on a brainstorming session

Transfected hipocampal neural cells with GFP

6.1 The Two-Photon Microscope

The development of the two-photon microscope in 1990 by Winfried Denk, James H. Strickler, and Watt W. Webb was driven by the need for *in vivo* imaging of deep tissue structures. They were the first to combine two-photon absorption with laser scanning confocal fluorescence microscopy. However, the theoretical foundation of two-photon absorption was established by Maria Göppert-Mayer in 1931 (Fig. 6.1). Two-photon absorption occurs when two photons, each with nearly double the wavelength and half the energy of continuous-wave laser light, are absorbed simultaneously to excite a fluorophore (Fig. 6.2). The process of generating excitation and emission in a two-photon system is complex, and modifications have been made to microscopes to produce a signal from a single focal plane (Fig. 6.3) without the need to use a pinhole, a noticeable difference with single-photon microscopy. A pulsed laser emission is generated through specialized optics with negligible power loss. Currently, two-photon microscopy is a powerful tool for noninvasive, intravital imaging of intact tissues or live animals. It offers improved penetration depth and optical sectioning, significantly reducing photodamage to tissues and photobleaching of fluorochromes.

Fig. 6.1 (continued) J. Hans D. Jensen, she published *Elementary Theory of Nuclear Shell Structure*. In 1963, they were awarded the Nobel Prize in Physics, along with Eugene P. Wigner, for their contributions to the understanding of nuclear shell structure

In the final years of her life, Maria Goeppert Mayer was a professor at the University of California, San Diego, and became a member of both the National Academy of Science and the Akademie der Wissenschaften, based in Heidelberg, Germany. She passed away on February 20, 1972

6.1 The Two-Photon Microscope

Fig. 6.1 Maria Goeppert Mayer
Goeppert Mayer was born on June 28, 1906, in Kattowitz, Upper Silesia, which back then was part of Germany. During these years, it was uncommon for women to attend university; nevertheless, Maria had always been determined to do so. In 1924, she began her studies at the Georg-August University with the intention of becoming a mathematician. However, she found herself increasingly drawn to physics, the area of study that she would continue her graduate studies
In 1930, Maria completed her doctorate in theoretical physics. Her dissertation investigated the theoretical impact of a high concentration of photons in space and time on their collective absorption. This work was groundbreaking in the fields of laser science and nonlinear optics. As we will discuss in this chapter, Goeppert Mayer's research was pioneering. Her work was honored by naming the unit of two-photon absorption probability after her. It was not until 1961 that the first experimental demonstration of fluorescence in calcium fluoride crystals confirmed the predictions made in her doctoral dissertation. In the 1990s, W. Denk, J. Strickler, and W. Webb constructed the first two-photon microscope, utilizing nonlinear effects such as two-photon fluorescence
In the summer of 1930, Goeppert Mayer returned to Göttingen, where she worked with Max Born before moving to the United States, embarking on a long career that would eventually lead to the Nobel Prize. During the Great Depression, she gave free lectures at Johns Hopkins University in Baltimore. In 1939, she worked on the electronic structure of transuranium elements and on the electronic structure of the atom at Columbia University in New York City. During World War II, she was invited to join a secret team working on the Manhattan Project, which resulted in the development of the atomic bomb
After World War II, in 1946, Maria became an associate professor at the Institute for Nuclear Studies at the University of Chicago. Later, she was promoted to professor and began researching nuclear physics. Goeppert Mayer established the significant strength of nucleon–nucleon interactions and explained how their movement was influenced by spin–orbit coupling. In 1955, alongside

The bigbang cells

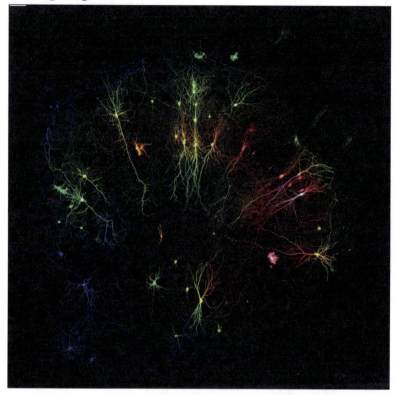

Transfected hipocampal neural cells with GFP

Fig. 6.2 (continued) wavelengths (and thus at lower energy). Due to this characteristic, two pulses are required to reach the "on state" compared to the single pulse of a CW laser (above). Additionally, in the "off state" of the two-photon system, as electrons return to their original orbitals, they are again reached by two pulses, generating emission with the same intensity as in single-photon systems (middle). The precise coincidence needed for an electron to be irradiated by two laser pulses is not trivial and requires specific conditions (detailed in Fig. 6.3). The lower diagram of this image is a detailed representation of the two-photon principle, where two pulses (represented by ½ in the figure) strike an electron to raise it to a higher energy state. Once the "on state" is achieved, the electron tends to return to its ground state, releasing the gained energy as light emission. At this emission moment, the electron is again reached by two Ti:Sa pulses, achieving stimulated emission by radiation and, consequently, the amplification of light

6.1 The Two-Photon Microscope 103

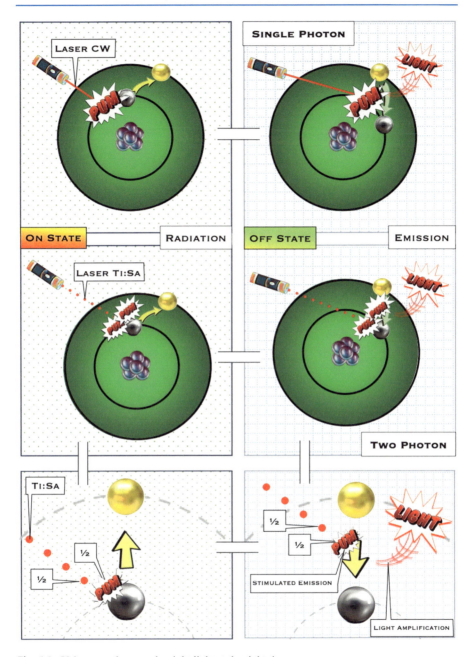

Fig. 6.2 Using two photons: the right light at the right time
Single-photon confocal microscopy uses a continuous-wave (CW) laser, which results in a constant energy flow that irradiates both the sample and the fluorochromes. This phenomenon, known as photodamage or phototoxicity, causes irreversible alterations to the fluorochromes and the sample. One of the differences between single-photon and two-photon systems is that the latter minimizes damage by using a pulsed, tunable titanium–sapphire (Ti:Sa) laser, which operates at longer

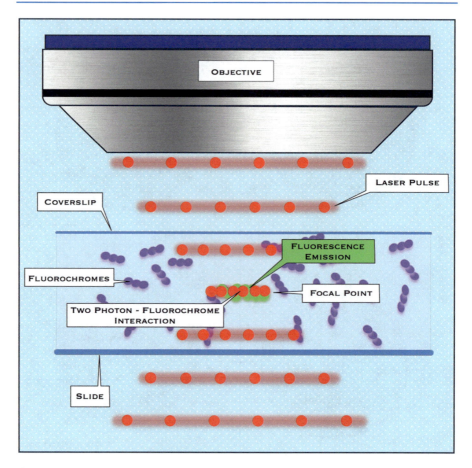

Fig. 6.3 Synchronicity: a perfectly timed instant of coincidence for two photons
An increased number of photons and the proximity between them in a sample raises the probability that two of them will simultaneously collide with a fluorochrome, making this moment more than just a coincidence. The interaction of two photons is a critical event in the two-photon system, as together they provide enough energy to excite a fluorochrome and subsequently induce the fluorophore emission. This primarily occurs in a confined area called the focal point. Above and below this zone, the distance between photons increases significantly due to a larger distribution area, which drastically reduces or nullifies the probability of photon synchronization and the subsequent excitation and emission

6.2 Infrared Light

Two-photon microscopy became feasible with the advent of high-power ultrafast-pulsed lasers. A mode-locked titanium–sapphire laser, which operates at an 80 MHz repetition rate with pulse durations of around 100 femtoseconds and offers tunable pulses in the mid to far infrared range (690–1090 nm), is particularly well suited for this application. However, titanium–sapphire lasers are less effective at wavelengths beyond 1000 nm. For these longer wavelengths, alternatives such as ytterbium-doped lasers (1030–1060 nm) and chromium–forsterite lasers (1200–1300 nm) provide better performance. Tunable infrared lasers also enable the excitation of a wide range of fluorophores commonly used in biological research. As infrared light has a longer wavelength, nearly double the wavelength is required for two-photon absorption to achieve the same energy as shorter wavelengths. For example, in single-photon microscopy, green fluorescent protein (GFP) is excited at 488 nm and emits at 550 nm. In two-photon microscopy, GFP is excited at 900 nm but still emits at 550 nm. Therefore, although the excitation wavelength in two-photon microscopy is nearly double that in single-photon microscopy, the emission wavelength remains unchanged.

Several adaptations have been made to two-photon microscopy to improve the efficiency of two-photon absorption, which naturally has a low probability of occurrence. To achieve this, two-photon microscopy uses high numerical aperture objectives to increase the spatial concentration of photons (Fig. 6.4). Generally, the laser beam is scanned across the sample using an XY-galvanometric mirror (galvo), enabling raster scanning of each location and constructing images on a pixel-by-pixel basis.

Additionally, highly sensitive detectors such as photomultiplier tubes (PMTs) have been developed to optimize several factors, including large photosensitive areas (on the millimeter scale), quantum efficiency, gain, absorption spectra, and dark noise. A typical two-photon microscope configuration, therefore, includes a titanium–sapphire pulse laser, galvanometer mirrors, larger numerical apertures objectives, and highly sensitive PMTs (Fig. 6.5). During operation, the pulsed laser spot is scanned across the sample by the galvanometers (as described in Chap. 5), fluorescence is collected and directed onto the PMT, and the resulting signal is recorded and converted into an image using imaging software.

6.3 Deep Imaging

A key feature of two-photon microscopy is its ability to image at substantial depths, reaching up to 1 mm depending on the properties of the tissue. Infrared light is scattered more than ten times less than visible light, making it particularly advantageous for deep tissue imaging. Scattering is the primary factor limiting imaging depth; as depth increases, fewer incident photons are able to reach the focal point. To counteract scattering, higher laser power must be applied to the sample. With a conventional titanium–sapphire laser, imaging depths of approximately 800 μm can be achieved in the brain cortex of an adult mouse. However, new techniques and modifications, such as using alternative light sources, are being explored, as longer wavelengths can penetrate tissue more effectively.

Other alternatives include using microendoscopes or removing overlying structures to access deeper tissues. Despite its limitations, the development of two-photon microscopy has been a groundbreaking advancement, especially in neuroscience. Recently, its applications have expanded to a variety of other fields,

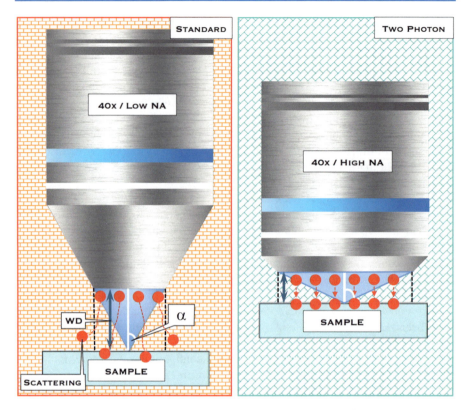

Fig. 6.4 Photon herding
Unlike a single-photon microscope, the multiphoton microscope uses objectives with larger numerical apertures (NA). This increase implies a larger diameter of the front objective lens, an adaptation that allows for more photons from the infrared laser to pass through the lens and reach the sample. As a result, the probability of two photons colliding at the focal point increases significantly. In other words, a larger numerical aperture improves the collection of a higher number of photons due to two main factors: (a) The reduction of the distance between the front objective lens and the sample (working distance [WD]). The closer proximity between these two elements reduces photon loss by minimizing the chance of photon scattering, as there is less likelihood of deviation in their trajectory. (b) The increase in the dimensions of the illumination cone, which results in a greater number of photons being directed toward the sample. For more details on alpha angle, numerical aperture, working distance, and objective color code, refer to Chap. 3

including immunology, cancer research, plant biology, stem cell research, and angiology.

6.4　In Vivo Imaging

The primary goal of developing two-photon microscopy was to enable *in vivo* imaging. This objective has driven significant advancements in the development of fluorescent proteins, a field that continues to evolve. Currently, many tools are available,

6.4 In Vivo Imaging

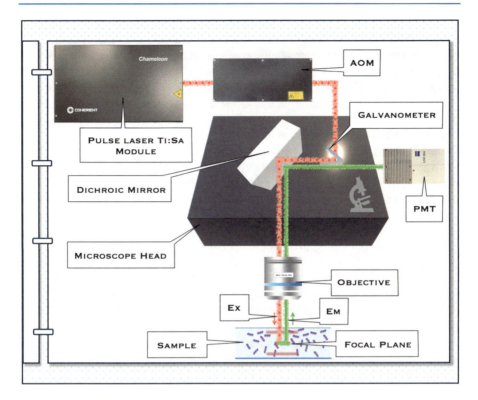

Fig. 6.5 Achieving efficiency and precision
This diagram provides a general representation of the two-photon microscopy system. The process begins with the infrared laser (Ti:Sa), which emits pulsed laser beams; this would mean the light is of high frequency and not continuous. The pulses would then reach an acousto-optic modulator (AOM) or a Pockels system, both of which serve to reduce the initial energy potential of the infrared laser. The initial energy of the pulsed laser before entering the AOM is on the order of watts, but after passing through the system, it is reduced to the milliwatt range. Once the laser is attenuated, it is reflected by a dichroic mirror that redirects the light beam to the sample. The dichroic mirror, described in Chap. 4, is a selective element that reflects infrared wavelengths toward the galvanometer mirrors. These, in turn, direct and move the laser across the sample to excite the fluorochromes by scanning each point in the X and Y planes after passing through the objective lenses. At this point, the fluorescence emitted by the fluorochromes at the focal plane returns to the objective lenses and is then directed back to the dichroic mirror, which now functions as a polarizing filter, allowing the fluorescent emission from the sample to pass through. This light reaches the highly sensitive photomultiplier tube (PMT) system, where the light information is amplified. The PMT is the final relay in the system before the information is communicated to a computer system

including genetically encoded calcium indicators, voltage indicators, mitochondrial calcium sensors, uncaging compounds, and opsins. These molecules are crucial for recording neuronal activity, modulating synaptic function, and controlling activity with light *in vivo*, whether in awake animals or live cells.

Many brain studies make use of the previously mentioned molecules. Experiments conducted using two-photon microscopy often involve recording calcium signals across various brain regions in awake animals as they perform motor, visual, or sensory tasks over extended periods, sometimes spanning several weeks. Advances in imaging technology have greatly improved the precision of calcium signal recording, with galvanometers and resonant scanners now operating at frequencies between 5 and 30 Hz. Additionally, calcium sensors with varying decay times—slow, medium, and fast—are available to meet specific experimental requirements. These advancements illustrate how progress in imaging technology is closely aligned with molecular developments.

Another notable feature of two-photon microscopy is its high resolution, which allows for the imaging of very small structures, such as dendritic spines (approximately 1 μm in diameter), due to reduced scattering (Fig. 6.6). Besides terminal axons, dendritic spines are the primary sites of neuronal connections in the brain, known as synapses, with approximately 90% of synapses occurring at these structures. Changes in dendritic spines are closely linked to a specific neuronal activity. Prior to the development of two-photon microscopy, studying the dynamics of dendritic spines was challenging. Now, two-photon microscopy enables the imaging of dendritic spines in live neurons expressing GFP (Fig. 6.6, right bottom) and allows for modulation of their activity through two-photon uncaging of glutamate. With two-photon, the laser spot is approximately 1 μm^3, enabling targeted uncaging of glutamate (720 nm) with light pulses at individual dendritic spines, enhancing or inhibiting responses, while simultaneously imaging structural changes GFP-positive neurons (900 nm excitation).

6.4 In Vivo Imaging

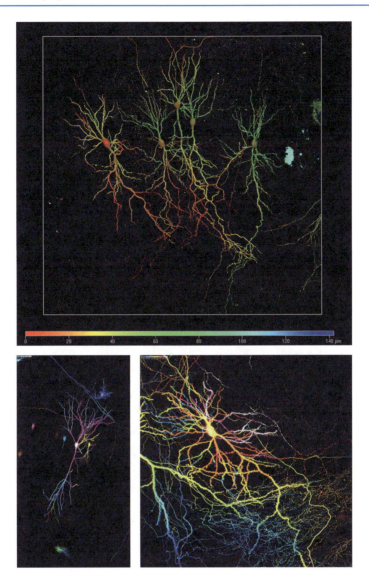

Fig. 6.6 The inclusive neurons
These are two-photon micrographs showing the Z-projections of GFP-positive hippocampal pyramidal neurons acquired from an organotypic slice culture. The color pattern represents the depth of the cells (as below in the 3D reconstruction image, on top), with cool colors indicating deeper areas and warm colors representing more superficial areas. Through these beautiful images, it can be appreciated how the hippocampal structure is preserved and how the neurons remain distributed throughout the thickness of the slices, in addition to showing the morphology and distribution of its processes, as well as the presence of spines on them (right bottom)

The moonwalk cells

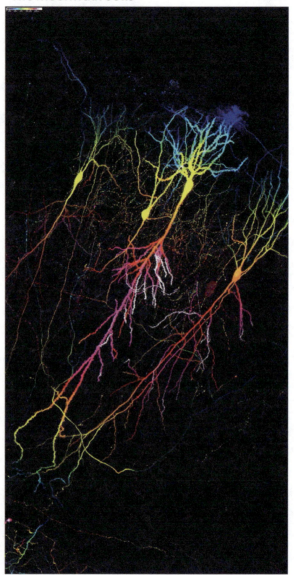

Transfected hipocampal neural cells with GFP

Most studies involving live animals are conducted with the animals either head-fixed or anesthetized. However, as the questions in neuroscience become increasingly complex, there is a growing need to study animals in natural, moving conditions and to record activity over larger areas (in the order of millimeters). Recent advancements have led to the development of miniature two-photon microscopes, which enable imaging large areas while animals move freely or engage in tasks. Alongside these advances, improvements in molecular tools—such as enhanced calcium sensors, voltage sensors, and opsins—have significantly progressed neuroscience research. It is now possible to record the activity of thousands of neurons simultaneously while the animal performs a task.

A two-photon microscope is not limited to imaging fluorescent specimens; since the two-photon is a nonlinear optical phenomenon, it can produce a second harmonic generation (SHG). SHG occurs in materials with non-centrosymmetric structures, where two photons interact with the material and are subsequently emitted as a single photon with double the original energy (or half the wavelength). Biological materials capable of producing SHG images include collagen, microtubules, myosin, and starch granules, all of which can be imaged without staining. Two-photon microscopy enables the generation of three-dimensional SHG images. This technique is increasingly valuable in cancer research, as tumors often alter collagen structure. SHG imaging is therefore ideal for studying intact biopsies and can expedite clinical results. Overall, two-photon microscopy has become an essential tool in neuroscience and biological research due to its numerous advantages.

The only drawbacks of this type of microscopy are its cost and the difficulties that come with hiring competent experts to operate it. In reality, super-resolution microscopy has no significant disadvantages; in fact, it can even be achieved using a modified two-photon microscope. Today, it is possible to record thousands of neurons in vivo using two-photon microscopy in conjunction with a machine learning interface (BMI). This advancement allows researchers to decode neuronal information while experimental subjects engage in tasks or even when they recall memories.

6.5   When Should a Two-Photon Microscope Be Used?

This system is recommended for imaging high-dimensional samples containing fluorochrome-labeled molecules that require reconstruction through the stacking of multiple optical planes along the axial axis. It is widely used for recording structures and signals in tissues exceeding the micrometer scale or for analyzing organs in small animals, both in fixed conditions and *in vivo*.

Two-photon microscopy allows for deep tissue penetration (exceeding 1 millimeter) and operates within a broad excitation range at the infrared spectrum. Its applications span from basic science to clinical settings, where it is commonly used for noninvasive optical techniques and for the pathological diagnosis of diseases such as cancer and immune disorders due to its ability to generate second harmonic light.

Due to the high sensitivity of its photomultipliers, it is an ideal option for analyzing thin samples with low fluorescence emission. Additionally, the use of a pulsed laser containing photons with half of the energy needed to achieve the emission of fluorophores reduces the energy absorbed by the fluorochrome, making it an excellent choice for imaging highly unstable fluorochromes. It also extends the lifespan of fluorochromes when performing long-term time-lapse imaging.

Further Reading

For two photon confocal microscopy basis: Chen et al. (2002), Denk et al. (1990), Helmchen & Denk (2005), Hira (2024), Svoboda & Yasuda (2006), Tsutsumi et al. (2024).

For second-harmonic generation: Aghigh et al. (2023), Hsiao et al. (2021), Stocker et al. (2023).

For two photon confocal microscopy techniques: Chen et al. (2013), Ellis-Davies (2019), Harvey & Svoboda (2007), Johnstone et al. (2019), Klausen & Blanchard-Desce (2021), Li et al. (2020), Liu et al. (2022), Marymonchyk et al. (2021), Mizuta (2021), Ozulumba et al. (2023), Piant et al. (2016), Pinkard et al. (2021), Reyes (2019), Trautmann et al. (2021), Zong et al. (2021), Zong et al. (2022).

For laser for two photon microscopy: Deguil et al. (2004).

References

Aghigh A, Bancelin S, Rivard M, Pinsard M, Ibrahim H, Légaré F (2023) Second harmonic generation microscopy: a powerful tool for bio-imaging. Biophys Rev 15(1):43–70

Chen IH, Chu SW, Sun CK, Cheng PC, Lin BL (2002) Wavelength dependent damage in biological multi-photon confocal microscopy: a micro-spectroscopic comparison between femtosecond Ti: sapphire and Cr: forsterite laser sources. Opt Quant Electron 34:1251–1266

Chen TW, Wardill TJ, Sun Y, Pulver SR, Renninger SL, Baohan A, Schreiter ER, Kerr RA, Orger MB, Jayaraman V, Looger LL, Svoboda K, Kim DS (2013) Ultrasensitive fluorescent proteins for imaging neuronal activity. Nature 499(7458):295–300. https://doi.org/10.1038/nature12354

Deguil N, Mottay E, Salin F, Legros P, Choquet D (2004) Novel diode-pumped infrared tunable laser system for multi-photon microscopy. Microsc Res Tech 63(1):23–26

Denk W, Strickler JH, Webb WW (1990) Two-photon laser scanning fluorescence microscopy. Science 248(4951):73–76. https://doi.org/10.1126/science.2321027

Ellis-Davies GC (2019) Two-photon uncaging of glutamate. Front Synaptic Neurosci 10:48

Harvey CD, Svoboda K (2007) Locally dynamic synaptic learning rules in pyramidal neuron dendrites. Nature 450(7173):1195–1200. https://doi.org/10.1038/nature06416

Helmchen F, Denk W (2005) Deep tissue two-photon microscopy. Nat Methods 2(12):932–940

Hira R (2024) Closed-loop experiments and brain machine interfaces with multiphoton microscopy. Neurophotonics 11(3):033405

Hsiao CY, Teng X, Su TH, Lee PH, Kao JH, Huang KW (2021) Improved second harmonic generation and two-photon excitation fluorescence microscopy-based quantitative assessments of liver fibrosis through auto-correction and optimal sampling. Quant Imaging Med Surg 11(1):351

Johnstone GE, Cairns GS, Patton BR (2019) Nanodiamonds enable adaptive-optics enhanced, super-resolution, two-photon excitation microscopy. R Soc Open Sci 6(7):190589

Klausen M, Blanchard-Desce M (2021) Two-photon uncaging of bioactive compounds: starter guide to an efficient IR light switch. J Photochem Photobiol C: Photochem Rev 48:100423

Li H, Yan M, Yu J, Xu Q, Xia X, Liao J, Zheng W (2020) In vivo identification of arteries and veins using two-photon excitation elastin autofluorescence. J Anat 236(1):171–179

References

Liu Z, Lu X, Villette V, Gou Y, Colbert KL, Lai S, Guan S, Land MA, Lee J, Assefa T, Zollinger DR, Korympidou MM, Vlasits AL, Pang MM, Su S, Cai C, Froudarakis E, Zhou N, Patel SS, Smith CL, Ayon A, Bizouard P, Bradley J, Franke K, Clandinin TR, Giovannucci A, Tolias AS, Reimer J, Dieudonné S, St-Pierre F (2022) Sustained deep-tissue voltage recording using a fast indicator evolved for two-photon microscopy. Cell 185(18):3408–3425

Marymonchyk A, Malvaut S, Saghatelyan A (2021) In vivo live imaging of postnatal neural stem cells. Development 148(18):dev199778

Mizuta Y (2021) Advances in two-photon imaging in plants. Plant Cell Physiol 62(8):1224–1230

Ozulumba T, Montalbine AN, Ortiz-Cardenas JE, Pompano RR (2023) New tools for immunologists: models of lymph node function from cells to tissues. Front Immunol 2023(14):1183286

Piant S, Bolze F, Specht A (2016) Two-photon uncaging, from neuroscience to materials. Opt Mater Express 6(5):1679–1691

Pinkard H, Baghdassarian H, Mujal A, Roberts E, Hu KH, Friedman DH, Malenica I, Shagam T, Fries A, Corbin K, Krummel MF, Waller L (2021) Learned adaptive multiphoton illumination microscopy for large-scale immune response imaging. Nat Commun 12(1):1916

Reyes AM (2019) Using two-photon microscopy to analyze Arabidopsis Thaliana leaves under autophagy conditions and second harmonic generation in Collagen. The University of Texas, El Paso

Stocker M, Baumeister P, Canis M, Vogel M, Gires O (2023) Second harmonic generation imaging of head and neck squamous cell carcinoma. Front Imaging 2:1133311

Svoboda K, Yasuda R (2006) Principles of two-photon excitation microscopy and its applications to neuroscience. Neuron 50(6):823–839

Trautmann EM, O'Shea DJ, Sun X, Marshel JH, Crow A, Hsueh B, Vesuna S, Cofer L, Bohner G, Allen W, Kauvar I, Quirin S, MacDougall M, Chen Y, Whitmire MP, Ramakrishnan C, Sahani M, Seidemann E, Ryu SI, Deisseroth K, Shenoy KV (2021) Dendritic calcium signals in rhesus macaque motor cortex drive an optical brain-computer interface. Nat Commun 12(1):3689

Tsutsumi M, Ishii H, Nemoto T (2024) Development of two-photon super-resolution microscopy. Brain Nerve 76(7):807–812

Zong W, Wu R, Chen S, Wu J, Wang H, Zhao Z, Chen G, Tu R, Wu D, Hu Y, Xu Y, Wang Y, Duan Z, Wu H, Zhang Y, Zhang J, Wang A, Chen L, Cheng H (2021) Miniature two-photon microscopy for enlarged field-of-view, multi-plane and long-term brain imaging. Nat Methods 18(1):46–49

Zong W, Obenhaus HA, Skytøen ER, Eneqvist H, de Jong NL, Vale R, Jorge MR, Moser MB, Moser EI (2022) Large-scale two-photon calcium imaging in freely moving mice. Cell 185(7):1240–1256

The Scanning Electron Microscope

7

The electron can no longer be conceived as a single, small granule of electricity; it must be associated with a wave, and this wave is no myth; its wavelength can be measured, and its interferences predicted.

Louis de Broglie

© The Author(s), under exclusive license to Springer Nature Switzerland AG 2025
A. Rosas-Arellano et al., *Microscopic Wonders*,
https://doi.org/10.1007/978-3-031-92559-7_7

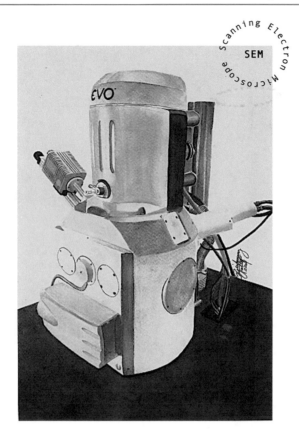

Before describing the scanning electron microscope, it is essential to know the difference between magnification and amplification in terms of resolution given the relevance acquired by electron light sources around this concept. In wide terms, *magnification* can be defined as the non-distorted increase in the size of an object using one or more lenses. In microscopy, this would be the resulting image given by the interaction between the microscope lenses, the light-emitting resource, and the sample. In theory, the more you increase the lens's magnifying power and the lighting resource's energy power, the higher the resolution; therefore, magnification is closely related to resolution. On the other hand, *amplification* refers to a procedure that does not involve the use of lenses, it is a digital increase of the size of a photomicrograph without implying the increase of the number of pixels composing it. Consequently, amplification is not directly linked to resolving power (Fig. 7.1).

The resolving power of a photonic microscope was first described by Ernest Abbe using the equation of the minimum distance between two points (Fig. 7.2). By applying Abbe's equation, we can get to know the resolution limit of an optical microscope. This, of course, implies that there is a resolution limit for microscopes

7 The Scanning Electron Microscope

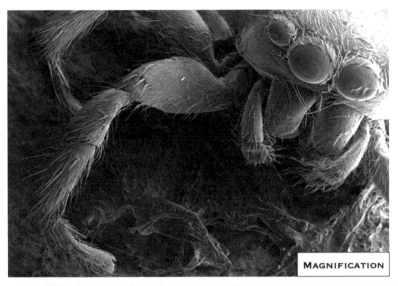

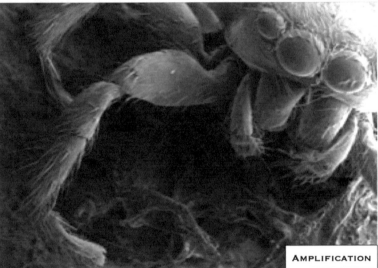

Fig. 7.1 Magnification vs amplification
In microscopy, magnification is referred to as the result of the interaction of the microscope's lenses with the light and the sample. What is called "total magnification" involves the eyepiece magnification used in the objective. For instance, if you use a target 60× magnification in combination with a 10× eyepiece, the final magnification would be 600X. The image in the upper panel has a size of 4096 × 3072 pixels, while the lower panel shows an image with a size of 208 × 154 pixels. Both images have a magnification of 30×; however, the second photomicrograph (lower one) was digitally amplified to have the same dimensions (14 cm long × 9.8 cm wide) as the top micrograph. This is a clear illustration of the differences between magnification and amplification

Fig. 7.2 Abbe and the minimum distance between two points equation

Ernest Abbe was born on January 23, 1840, in Eisenach, Germany. It has been noted that he displayed signs of unusual intelligence from a young age. When he was still a teenager, Abbe began his university studies in Jena and later in Göttingen, where he obtained his doctorate. He became a professor at the University of Jena, and in 1866, a pivotal moment in his scientific career occurred when he met Carl Zeiss. Abbe was the first to recognize the limitations of microscope magnification, and he modified the shape and arrangement of the lenses to influence their physical properties. The modifications he made improved the control of light refraction and dispersion in a microscope made for Carl Zeiss's company. These innovations surpassed any microscope that existed at the time. The high sales of Zeiss microscopes resulting from these improvements made Abbe an equal partner in the company in 1876. From then on, the company's growth was exponential, and they moved on to design other optical devices besides microscopes. Abbe's contributions globalized the company, which continued to grow year after year, eventually becoming one of the best microscopy companies worldwide. Ernest Abbe died in 1905, and to this day, his theory on the limits of resolution, which determines the minimum distance (d) that must exist between two points to differentiate them as independent, is still used by many microscopists to understand the resolution limits of a microscope. This formula considers the wavelength of the light used (lambda), the numerical aperture of the objective (NA), and the refractive index of the medium (n); it is for fluorescence microscopy, plus the numerical aperture of the condenser in bright field microscopy (adding another n, hence 2n), multiplied by the sine of the angle of alpha (sin α). For details of NA, refractive index, and light wavelength, see Chap. 3

that have conventional lenses (physical lenses), and lighting resources, such as the photonic ones presented in the previous chapters. This limit existed because the illumination wavelength with the highest energy, combined with the numerical apertures of the objectives (see Fig. 3.10), reached its limit at some point in the history of microscopy. In other words, there was no higher energy light source, nor could the proximity between the lens and the sample be shortened, both to favor a higher resolution limit.

Making an informative parenthesis. Nowadays, resolution in microscopy has exceeded the previously established limits by Abbe's equation. The main factor was the modification and use of light amplification derived from lasers, which favored resolutions of less than 200 nm and managed to reach up to 1 nm. This led to the creation of a subfamily within the family of fluorescence microscopes, commonly known as super-resolution microscopes, that unlike electron microscopy which only works with in vitro samples, super-resolution can work in both in vitro and in vivo samples.

Long before the birth of super-resolution microscopy, the limiting resolution of photonic microscopes laid the foundations for the creation of electron microscopy. It was in 1924 that the French Louis de Broglie (Fig. 7.3), postulated diffraction as one of the characteristics of electrons. Let us remember that diffraction is inherent to the wave motion of photonic light, de Broglie suggested that electrons should behave like a photon. Electronic diffraction thus represented a lighting resource with greater energy that had to exceed the lengths of conventional waves used in microscopy at that time.

It took years to consolidate enough evidence to back up the wave–particle duality of electrons. It was in 1927 that C.J. Davisson and L.H. Germer showed the indisputable wave nature (Fig. 7.4) through an elegant and clear experiment that confirmed the wave–particle hypothesis that was originally postulated by the French

Fig. 7.3 Broglie and electron momentum
Louis-Victor de Broglie was born on August 15, 1892, in Dieppe, France. He earned a degree in history in 1910 and subsequently obtained a science degree in 1913. In 1924, at the Faculty of Sciences of the University of Paris, he presented his doctoral thesis "Researches on the Quantum Theory," which included significant findings about the wave–particle duality of electrons, which was later confirmed by Davisson and Germer. He was a member and permanent secretary of the French Academy of Sciences, a professor at the Faculty of Sciences of the University of Paris, and at the Henri Poincaré Institute. His studies earned him various recognitions such as the Henri Poincaré Medal, awarded to him twice, the Kalinga Prize awarded by UNESCO, the gold medal awarded by the French National Scientific Research Centre, and most notably, in 1929, the Swedish Academy of Sciences awarded him the Nobel Prize in Physics for his discovery of the wave nature of electrons. He was named an honorary doctor of the Universities of Warsaw, Bucharest, Athens, Lausanne, Quebec, and Brussels. He was a member of 18 science academies in Europe, India, and the United States. After an immeasurable intellectual legacy, Louis de Broglie died on March 19, 1987

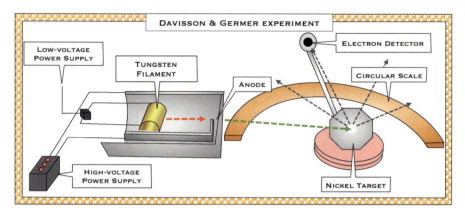

Fig. 7.4 Do electrons behave as waves?
Davisson and Germer designed an instrument to verify the dual nature of electrons. This vacuum system contained a high-voltage source and a low-voltage power supply that provided current to a tungsten filament. Released electrons (red dotted arrow) from the filament were attracted and accelerated by an opposite charge through an anode and directed (green dotted arrow) until they collided with a nickel metal surface. This surface reflected the electrons (greay dotted arrows), which were then attracted to a detector. The detector rotated to determine the angular dependence of the electrons scattered through the nickel surface, which means that the system determined where the electrons were most concentrated. Davisson and Germer found that there were different intensities or acceleration voltages of electron light at the various angles at which the detector rotated. This observation is consistent with the results of the wave nature of light from the double-slit experiment (Fig. 2.3); it was a clear demonstration that electrons behave like waves and thus convincingly demonstrated the wave–particle duality

physicist, who would later be awarded the Nobel Prize, for his postulate that electrons have dual nature, therefore they can be a source of illumination.

Simultaneously, in 1926, around 1 year before the Davidson and Germer experiment, the German Hans Busch developed the theory of the effect of magnetic fields on electrons, which would become pivotal in the birth of electron optics. An electromagnetic lens essentially functions like a conventional or physical lens and therefore follows the lens equation. This equation determines the distance at which a focal point and an image are formed based on a lens coupled with a light source (Fig. 7.5). In other words, the focal distance through a physical and an electromagnetic lens is predictable using this equation.

Combined, both elements: the electron source and the electromagnetic lenses, laid the foundation for the development of the first transmission electron microscope, which will be discussed in the next chapter. On the other hand, the first scanning electron microscope was developed by the German scientist Von Ardenne in 1938 (Fig. 7.6), is estimated that the resolution of this scanning electron microscope was 100 micrometers. The type of information provided by this type of microscope is surface-related, which would be similar to the images generated by the stereoscopic binocular microscope described in Chap. 3 (Fig. 3.12), both microscopes use mainly reflective information.

7 The Scanning Electron Microscope

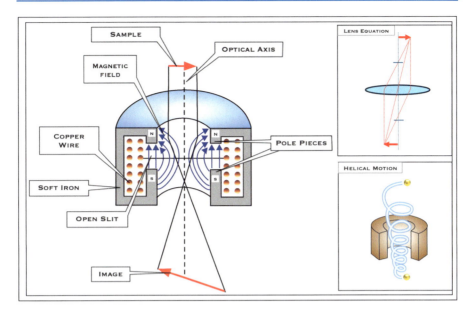

Fig. 7.5 Electromagnetic lenses: ingenuity and precision
Just as in a conventional microscope, where lenses control the passage of photons, in an electron microscope, electromagnetic lenses are used to control the ordered flow of electrons and concentrate them at a given point, which would be *the focal point*. The lenses for these microscopes must be electromagnetic because it is improbable that electrons will pass through a physical lens. In the attached diagram, we can see that they are composed of coils that contain copper wires; these coils are coated with soft iron to increase the electromagnetic field. When a current is applied to the coils, the electromagnetic field is generated thanks to the slits located in the inner area of the lenses, which provide polarity. This magnetic field attracts the electrons, and the speed and quantity of electrons can be controlled by the current supplied to the coils. The passage of electrons through the electromagnetic lenses creates a focal point, which follows the equation of physical lenses (above in the right-side figure). Relevantly, the passage of electrons from one side to the other of the electromagnetic lens occurs in a helical (right, below) and not in a linear motion, as would be the case for photons crossing a glass lens

A scanning electron microscope, also known as SEM (due to its acronym), relies on a high voltage that supplies current to a tungsten filament located just below the high-energy connection located at the top of this microscope. The voltage–filament interaction generates a plethora of electrons, this component is designated as the cathode, which has a negative charge. Subsequently, the electrons are attracted and accelerated downwards by an element with a positive charge, the anode. The electrons will travel toward the sample through the central hole of a cylinder called a Wehnelt cylinder, named after the physicist Arthur Rudolph Berthold Wehnelt, who designed this cylinder to focus the electrons and dissipate energy. This energy-generating and routing area is called the "electron gun" (Fig. 7.7). In summary, an electron gun has three components: (1) a filament or cathode, (2) a Wehnelt cylinder, and (3) an anode.

Fig. 7.6 The emergence of scanning microscopy
Manfred Von Ardenne, a German physicist, was born in Hamburg on January 20, 1907. In 1937, after the invention of the transmission electron microscope, Von Ardenne created the first scanning microscope. As a man with outstanding versatility, his was a distinguished career full of scientific and academic achievements, as well as invention patents. He contributed to the medical and nuclear fields, as well as to the development of radio and television, Manfred Von Ardenne died on May 26, 1997, in Dresden. The first commercial scanning electron microscope (SEM), which was based on Manfred's design, had a resolution of around 10 nm. Nowadays, SEMs can achieve up to 1-nm resolution using an energy of 30 keV

7 The Scanning Electron Microscope

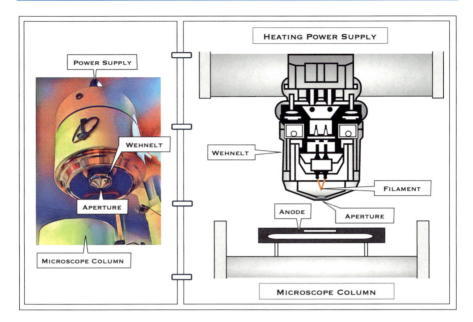

Fig. 7.7 The electron gun, a piece of ingenious tech
The electron gun (left) consists of a filament (a cathode), a Wehnelt cylinder, and an anode (right). This component of the SEM uses a high voltage of up to 30,000 volts, which is applied to a filament. When subjected to an electric charge, the filament generates electrons, resulting in a negatively charged cloud beneath it, which will be focused or condensed through a central aperture in the cylinder. The anode participates in attracting the electrons to direct them out of the electron gun toward the lower part of the microscope. The anode can be biased or charged with higher or lower energy to accelerate a greater or lesser number of electrons, respectively, as appropriate for each situation

The spirulina maxima, a little factory of oxygen

The current of the electron beam supplied between the filament and the Wehnelt cylinder can be regulated electronically by a resistance known as bias. Increasing the current is directly related to a higher flow of electrons that will collide with the sample at a higher speed. This will facilitate achieving greater electron penetration and an increase in the spot size, which means the diameter of the electron beam that hits the sample. A greater number of electrons interacting with the sample means more detailed information obtained of the observed sample.

Described in order from top to bottom inside an SEM, it can be found that below the electron gun the condenser electromagnetic lenses, which, as in photonic microscopes, have the function of linearly concentrating and aligning the light beam, but in this case the electron beam to pass through the microscope column until reaching the apertures' region. The apertures are a dynamic element of the SEM; they can be adjusted to coordinate with the magnification employed. Further down the microscope, the scan coil (or electromagnetic scanning lens) is located, just above the sample, this lens will direct the electron beam toward the sample once it has passed through the objective lens (Fig. 7.8).

7 The Scanning Electron Microscope

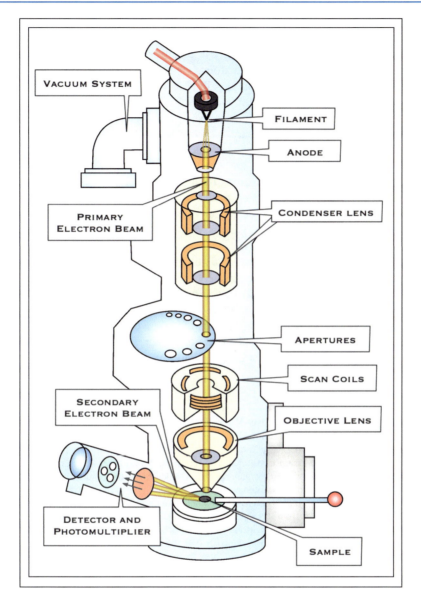

Fig. 7.8 Scanning electron microscope lenses
Below the electron gun is a system of electromagnetic lenses, which includes three main types: (1) the condenser lenses (usually composed of two of them), (2) the scanning lenses, and (3) the objective lens. The condenser lens, as its name implies, has the function of concentrating the electron beam on its way to the sample. This lens, in combination with the aperture, concentrates and dictates the thickness of the electron beam. Therefore, both elements play a fundamental role in resolution; smaller apertures are recommended for higher magnification, while larger ones are recommended for lower magnifications as they illuminate wide areas of the sample (as observed in the lower part of the figure). The scanning lens, an element that gives this microscope its name, is used to deflect the primary electron beam and subsequently perform a point-by-point scan of the sample. Finally, the objective lens provides the optimal working distance to correctly focus the electron beam on the sample

When electrons collide with the surface of the sample, various phenomena are generated from the primary beam, also called the high-energy incident beam. The consequences of this interaction can be the following: backscattered, secondary beam, X-ray, absorption, Auger electrons, visible rays, diffraction, and in some cases where some areas of the sample allow it due to their limited thickness, the electrons can be transmitted through it (Fig. 7.9).

In electron dynamics, the secondary beam is the one that contains less energy, which favors the attraction of electrons by a detector. The collection of secondary electrons coming from the sample provides its topographical information. The depth of the reading will depend on factors such as the quantity of electrons and their acceleration (which are governed by the voltage and the spot size) attaining readings of several microns. Greater depth topographic readings will require higher energy than this measured in kiloelectron volts (keV), here the consistency of the sample must be considered, as organic samples will require less energy to make depth readings than inorganic ones.

Once the sample has emitted secondary electrons, an electron detector is used to integrate the collected information. An electron detector is composed of various elements, and its operation is coupled to the number of secondary electrons released by the sample that are close enough to be recorded by the machine. This component contains a positive charge to attract the electrons. The positive charges in the atmosphere are dynamic and controlled by another microscope component, the bias. Once attracted, the electrons enter the photomultiplier (Fig. 7.10).

Fig. 7.9 (continued) energy-dispersive X-ray spectroscope analysis system, also known as EDS. This system detects X-rays emitted by a sample to determine its chemical composition. The generation of light (white arrow), also known as cathodoluminescence, is the production of photon light within the visible spectrum of a material subjected to high-energy electron bombardment. This technique is mainly applied to inorganic samples to determine their mineral composition. On the other hand, diffracted and transmitted electrons (gray arrows) that move away from any detector will be directed to the ground; whereas, some others will absorb and dissipate as heat within the sample. Finally, Auger electrons (orange arrow), analyzed in SEM through scanning Auger microscopy, provide important information that will be detailed later

Usually, the secondary beam, having lower energy than the rest, is quickly attracted by the electron detector in the high vacuum mode. There are three main types of secondary beams (illustrated in the lower panel): (a) those that interact with the sample and contain energies lower than 50 eV. Of these, about 90% of them will contain less than 10 eV, and within this population, most will have energies between 2 and 5 eV, (b) the second path of secondary electron generation is by the interaction of primary beam electrons with backscattered electrons, and (c) the third is the result of the interaction of backscattered electrons with parts of the equipment found near the sample, which causes electrons to lose significant energy. Conversely, the backscattered electron beam (yellow arrow), which contains high energy and therefore cannot be recovered in a high vacuum, is the most important when working in low vacuum mode. In this mode, when the backscattered beam interacts with free molecules in the sample chamber, it loses substantial energy. Similarly, the secondary beam loses energy significantly, making the latter impossible to recover, while the former is recovered through a specialized detector. Within the sample, the regions where each electron beam is generated are represented with the color code indicated above

7 The Scanning Electron Microscope

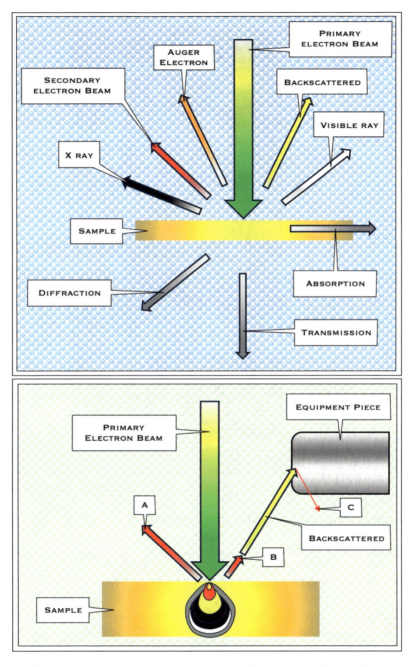

Fig. 7.9 The electron beam – sample interaction: a spread movement - Allegro molto
The interaction with the sample of the primary electron beam (green arrow) generates numerous consequences, the most important being the generation of the secondary electron beam (red arrow). Other reactions promote the generation of X-rays (black arrow) as a consequence of the electron bombardment that the sample receives is valuable when an SEM is equipped with an

The Climbing Arm of a House Cricket

After the photomultiplier, the image is formed. The image formed by an SEM microscope gives the impression of being three-dimensional; however, there is a slight deception here, which is that our brain is accustomed to interpreting combined tones of brightness and darkness as three-dimensional images.

The point-by-point reconstruction of the sample scan assigns values that will be represented in shades within a monochromatic scale. In this scale, an area of the sample that is close to the detector will be represented with lighter shades, while distant areas will be shown in darker shades (Fig. 7.11).

Besides the secondary electron beam, there are others that provide valuable information, such as Auger electrons. The high energy at which electrons collide with the elements composing the sample causes, on one hand, the displacement of electrons from a lower orbital to a higher energy orbital and, on the other hand, the expulsion of electrons from their respective atoms (Fig. 7.12).

The Auger effect is named in recognition of Pierre Victor Auger, who is credited with first describing this event. The "fingerprint" left by the departure of an electron after the high-energy absorption of an atom and its subsequent emission of light is

7 The Scanning Electron Microscope

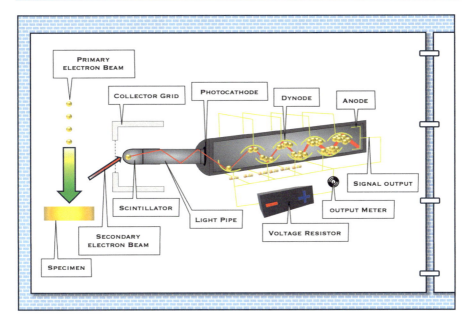

Fig. 7.10 Electron cloning
The capture of secondary electrons is accomplished by a scintillator-photomultiplier system. This attraction and subsequent multiplication of electrons aims to convert electrical information into digital. How does it happen? When an electron from the secondary beam is attracted by the detector, specifically by the scintillator located within the Faraday cage (collector grid), it is transported to the photomultiplier through a light pipe and converted by the photocathode into a photoelectron for entry into the photomultiplier body. In this system, just like in the photomultipliers of confocal microscopes, a photoelectron, through a cascade of events, is multiplied by a system of dynodes, which sequentially increase their charges to attract a greater number of electrons per stage. This process continues until the last dynode, which precedes the anode. The photoelectric signal then leaves the system to be converted into digital form, which is captured using a computational system to integrate the image

the basis of spectroscopy. Spectroscopy is a scientific discipline that provides analytical information about the ionic composition of a sample that we want to study and that is subjected to radiation. It generates information about the absorption and emission of specific energies by a single or a combination of chemical elements of the periodic table.

Independent of whether an analytical technique is used or not, an SEM operates using two basic systems: low vacuum and high vacuum. A vacuum internal environment allows electrons to move freely without interacting with molecules inside the microscope column. High vacuum involves the activation of vacuum pumps at two levels: (a) the microscope column and (b) the sample chamber. In contrast, low vacuum involves variable pressure and the deactivation of the vacuum, these programmable vacuum variations only occur within the sample chamber (Fig. 7.13). Low vacuum is recommended for samples that cannot withstand the pressure of

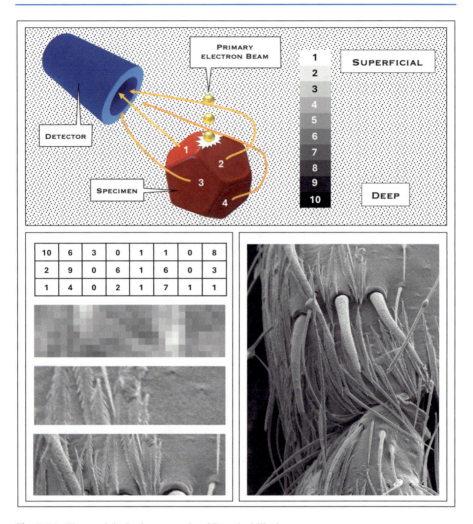

Fig. 7.11 The magician's show: creating 3D optical illusions

When we first observe an SEM photomicrograph, it gives us the impression of a projection with perfect three-dimensional characteristics. This misinterpretation of a 2D image is due to our stereoscopic vision, which we have adapted to a world full of relief. The information of objects illuminated by a light source in the outside world, through their shapes captured by our vision, is interpreted by our brain in the form of thickness, height, and depth to form a three-dimensional panorama. The dark tones of a landscape are perceived as areas of depth; in contrast, light tones are translated as surface areas. Importantly, the shades between light and dark tones represent intermediate planes that help integrate the interpretation of thickness, height, and depth to form a three-dimensional landscape

In an SEM, the point-by-point reconstruction of a sample considers the distance between the low-energy electron and the electron detector itself. The shorter the distance between them, the lower the value assigned in the formation of an image; conversely, electrons coming from a greater distance are assigned higher values, with intermediate values falling in between (top). Together, low, intermediate, and high values, when translated into a monochromatic scale for image formation, correspond to light tones, various shades of gray, and dark tones, respectively (bottom)

high vacuum. Examples of these would be soft biological tissues with little resistance or tension and high water content, or delicate and irreplaceable samples that cannot be subjected to critical point drying and gold coating.

In low vacuum settings, secondary electrons lose energy when interacting with free molecules present in the atmosphere of the sample chamber, making it difficult to recover this information using a conventional detector. Therefore, the use of a retractable auxiliary detector becomes necessary. This detector specializes in recovering backscattered electrons, which have higher energy (as mentioned previously) but have decreased it in the low vacuum.

One of the disadvantages of a low vacuum system could be the low resolution that arises from the scattering of the primary electron beams from its interaction with particles that do not belong to the sample and that are found free in the sample area, this results in a poor signal-to-noise ratio. Often, achieving high resolution in a low vacuum setting is closely related to the composition of the sample more energy it withstands, the better the image will be.

Finally, related to the sample, prior to placement within the SEM, it can be subjected to gold ionization, this procedure involves the use of a gold ionizer to create a vacuum atmosphere and then applying an electric discharge to a pure gold plate in a process known as thermionic ionization. The discharge heats the plate to the point of exceeding the binding energy of the gold ions, releasing them from the plate and generating plasma, which then bombards the sample ionically. Coating with gold, a biocompatible element chemically inert to electrons, aims to minimize absorption and enhance the reflection of electrons, including in areas with fine and thin structures, where electrons might otherwise be transmitted instead of reflected (Fig. 7.14).

7.1 When Should a Scanning Electron Microscope (SEM) Be Used?

A scanning electron microscope is the best option for analyzing minute topographical details with depth of field in small or large areas of an organic or inorganic sample. Some of these details include (a) morphology, (b) structural density, and (c) chemical composition (if the SEM is equipped with energy-dispersive spectroscopy, EDS). Additionally, techniques such as immunogold labeling can be suitable when colloidal gold particles are located very close to or right at the surface of the sample (especially when larger gold particles are used).

The surface relief details observed with SEM are so precise that it is commonly used in forensic science. In this context, the imagery produced by the microscope can confirm or rule out whether a hair found at a crime scene matches a suspect's hair based on the unique surface patterns present in an individual's hair structure.

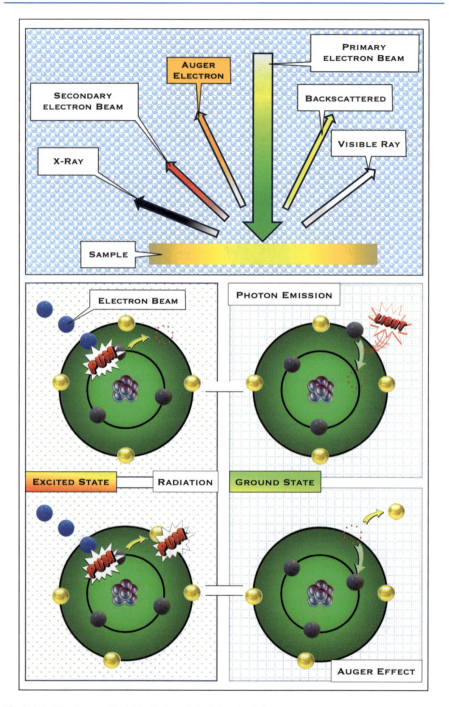

Fig. 7.12 The Auger effect: the fingerprint of chemical elements
The Auger electrons are generated by the interaction of the primary electron beam with the sample (on top). The scheme below illustrates a situation in which a high-acceleration and high-energy electron collides with an electron of an atom in a hypothetical sample. The incident energy

7.1 When Should a Scanning Electron Microscope (SEM) Be Used?

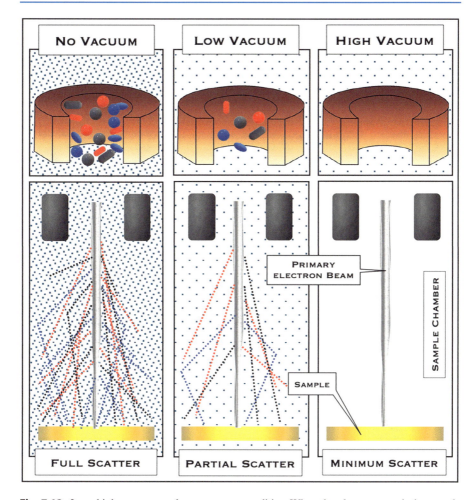

Fig. 7.13 Low, high vacuum, and non-vacuum conditions When the electrons reach the sample chamber in non-vacuum conditions (hypothetically), they could interact with a myriad of free molecules in it (left side of top and bottom figures). In low vacuum conditions, there will be partial dispersion (center), while a high vacuum system minimizes the interaction of the primary beam with free molecules in the chamber (right)

←

Fig. 7.12 (continued) galvanizes the electron to elevate to another energy level or orbital, leaving a vacancy in the orbital it previously occupied. Two events can occur in this vacant space: (1) that electron or another within the same orbital loses energy and returns to the vacant spot within the original orbital, releasing energy in the form of light along the way, this event is known as X-ray fluorescence (shown at the middle line of the figure); (2) alternatively, the electron that gains energy may collide with another electron in a higher orbital, causing the impacted electron to be ejected from the atom. This phenomenon is known as the Auger effect (bottom line of the figure). Scanning Auger microscopy provides information about the spatial distribution of chemical elements on the surface of the sample analyzed in a scanning electron microscope. X-ray fluorescence microscopy is one of the specialized functions of one of the most powerful microscopes today, the synchrotron

The Golden huntress

Fig. 7.14 The golden huntress

Gold has a fascinating role in the history of both ancient and modern civilizations and is recognized as one of the outstanding symbols of power and wealth. It has been considered a sacred, mystical, and healing element by countless ancient cultures that additionally associated gold with the sun due to its color and brilliance

Likewise, for science, gold is a prominent element. It is located in the group of transition metals. It has been assigned atomic number 79 within the periodic table where it is represented by the symbol Au, which is the abbreviation of the Latin "*Aureum*" meaning golden. Its classification corresponds to that of noble metals. This assignment is because, under regular environmental conditions, gold is chemically stable, stainless, incorrosible, and inert. The latter means that it reacts little or scarcely with other elements. All these physicochemical characteristics favor its wide use in the field of science and technology; without a doubt, it is one of the most popular elements

In microscopy, its use as a coating for SEM samples (as it was done on this microphotographed spider) reduces the charge given by the accumulation of electrons on the surface of the sample, which prevents a large number of pixels with white tones in an SEM micrography and consequently the loss of topographic information due to the high accumulation of electrons. An optimal gold coating favors images with good resolution and sharpness; this is because the reflection of electrons on the gold surface coating promotes the uniform and abundant emission of secondary electrons, while the absence of this coating would favor the absorption in the sample of electrons that make up the primary beam, considerably reducing the emission of secondary electrons

Further Reading

For electron microscopy basis: Ayache et al. (2010), Bozzola & Russell (1999), Dykstra & Reuss (2011), Hall (1953), Hall (1985), Harris (1991), Egerton (2005), Merchant (1994), Yacaman & Reyes (1995), Vázquez-Nin & Echeverría (2000), Williams & Carter (2009).

For scanning electron microscopy: Nguyen & Harbison (2017).

References

Ayache J., Beaunier L., Boumendil J., Ehret G., Laub D. 2010. Sample preparation handbook for transmission electron microscopy: techniques. Springer New York. XXV, 338

Bozzola JJ, Russell LD (1999) Electron microscopy: principles and techniques for biologists. Jones and Bartlett series in biology

Dykstra MJ, Reuss LE (2011) Biological electron microscopy: theory, techniques, and troubleshooting. Springer, New York. https://doi.org/10.1007/978-1-4419-9244-4

Egerton RF (2005) Physical principles of electron microscopy, vol 56. Springer, New York

Hall CE (1953) Introduction to electron microscopy. McGraw-Hill, Hightstown, p 451

Hall CE (1985) The beginnings of electron microscopy, advances in electronics and electron, vol 231. Physics Academic Press, New York, p 63

Harris JR (1991) Electron microscopy in biology. A practical approach. IRL Press/Oxford Univesity Press, Oxford, p 328

Merchant H (1994) El inicio de la microscopía electrónica en México. Ciencia y Desarrollo

Nguyen JNT, Harbison AM (2017) Scanning electron microscopy sample preparation and imaging. In: Espina V (ed) Molecular profiling. Methods in molecular biology, vol 1606. Humana Press, New York. https://doi.org/10.1007/978-1-4939-6990-6_5

Vázquez-Nin G, Echeverría O (2000) Introducción a la Microscopía Electrónica Aplicada a las Ciencias Biológicas. Facultad de Ciencias, UNAM. Fondo de Cultura Económica, p 163

Williams DB, Carter CB (2009) Transmission electron microscopy: a textbook for materials science. Micron 28(1):75–75

Yacaman MJ, Reyes J (1995) Microscopía Electrónica: Una visión del microcosmos. Consejo Nacional de Ciencia y Tecnología. Fondo de Cultura Económica, p 143

The Transmission Electron Microscope

8

The most important thing in science is to be together with a colleague at that moment when you discover something. When some little piece of nature unfolds, and you see something new.

John Heuser

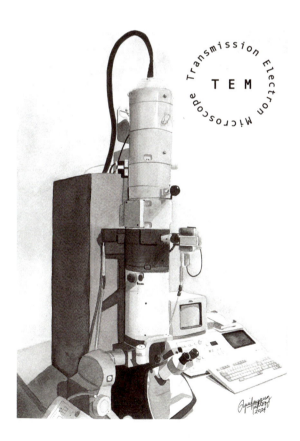

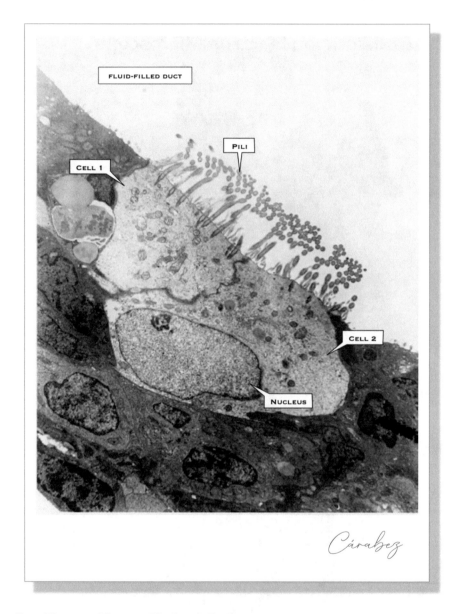

From Photons to Electrons: The Beauty Structure
For the very first image that we will dive into in this chapter, we have two adjacent cells, one of which shows very clearly the presence of a nucleus. The fascinating aspect of this picture lies in the elongated projections that come out of the cells. These projections are called "pili" or villi. They protrude from the cell membrane surface and are key for these cells to carry out their functions. The pili are oriented toward a duct filled with a liquid where solid structures, called otoliths, float. When the otoliths move around the liquid environment, they stimulate the villi, which in turn transmit signals back into the cell, and transduce information regarding the body's position in space.

The full appreciation of these structures is made possible thanks to the use of transmission electron microscopy.

8 The Transmission Electron Microscope

The invention of the transmission electron microscope (TEM), designed and manufactured during the 1930s, preceded the first historical appearance of the scanning electron microscope (SEM). Year after year since it was created, the TEM itself and its techniques have been distinguished for providing the highest resolution compared to any other type of contemporary microscope. Since some basic aspects of electron microscopy were previously covered when talking about the characteristics and principles of the SEM (Chap. 7), this chapter will only address those elements that make the TEM unique.

Even at first glance, the design of the first TEM prototype shows how its creator had notable scientific curiosity and an ingenious nature (Fig. 8.1). This microscope was the doctoral project of Ernest Ruska, who was under the mentorship of Max Knoll (Fig. 8.2). The first commercial piece was later developed by E. Ruska and von Borries, who was Ruska's peer during their postgraduate studies. The microscope they created achieved a resolution of 10 nm, surpassing any known instrument at the time. Nowadays, some modern TEMs can exceed a resolution threshold

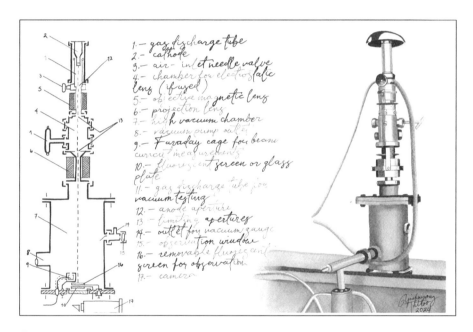

Fig. 8.1 The first transmission electron microscope
The first TEM was built by Ruska and Knoll in 1931 at the High Voltage Laboratory of the Technical University of Berlin. The paint on the right shows the external features of this instrument. Although rudimentary in appearance, the construction of each component is undoubtedly inspired. To be operated, this microscope required a minimum current of 30 keV and could take up to a maximum of 75 keV. With this energy, its resolving power surpassed any microscope that preceded it, marking the beginning of a new era: that of electron microscopy. The relevancy of this invention is of such magnitude, that when comparing new TEM models to any contemporary microscope from another category, no instrument exceeds its resolution. The diagram on the right shows the components of this prototype. These illustrations laid the groundwork for Ruska and his graduate colleague Bodo von Borries to produce the first commercial TEM for Siemens. Today, electron microscopy is recognized as one of the most important scientific advances of the twentieth century

Fig. 8.2 Max Knoll and Ernest Ruska
Max Knoll (pictured on the right) was born on July 17, 1897, in Germany. He was an electrical engineer from the University of Munich who later pursued his doctorate at the Institute for High Voltage Technology (Technical University of Berlin). In 1927, he worked as the leader of an electronics research group in which he collaborated with his student Ernest Ruska (picture on the left) to develop the first transmission electron microscope as part of Ruska's doctoral thesis. Ruska, also of German nationality, was born in Heidelberg in 1906. In 1931, he earned his degree in electronic engineering at the University of Munich. Over the years, Ruska continued to specialize in high voltage and high vacuum technology. Like Knoll, Ernest pursued his doctorate at the Technical University of Berlin in 1934, where he focused on the modification and shortening of the focal lengths of electron lenses originally designed by Hans Busch in 1926. These modified lenses, which for the first time included an iron casing, would be fundamental to the birth of transmission electron microscopy
Thanks to the intellectual contributions left by De Broglie, Davisson, Germer, and Busch, along with his own innovations in short focal length electromagnetic lenses, the design for the first electron microscope earned Ruska various prestigious accolades. The most important of these recognitions came in 1986 in the form of the Nobel Prize in Physics. Following a lifetime of continuous efforts to improve transmission electron microscopy and an invaluable intellectual legacy, Ernst Ruska passed away in 1988
After supervising Ruska's thesis, Max Knoll served from 1945 to 1966 as a professor at the University of Munich, a director of the Institute of Electromedicine, and at the Electrical Engineering Department at Princeton University in the United States. Max Knoll died on November 6, 1966, in Munich, Germany

of 50 picometers, that is, less than five-billionths of a centimeter, this is numerically expressed as 0.000000005 cm.

Externally, a TEM consists of the following structures: (a) a high-voltage source, (b) a column, (c) a projection or observation area, and (d) camera and image recording controls and commands (Fig. 8.3, left panel). In general terms, a TEM operates following the same principle as a transmitted light microscope with an inverted microscope configuration. Therefore, both essentially have the same components: (a) light source, (b) condenser, (c) specimen holder, (d) objectives, and (e) eyepiece. For the TEM, the latter corresponds to the projection lens (Fig. 8.3, right panel).

8 The Transmission Electron Microscope

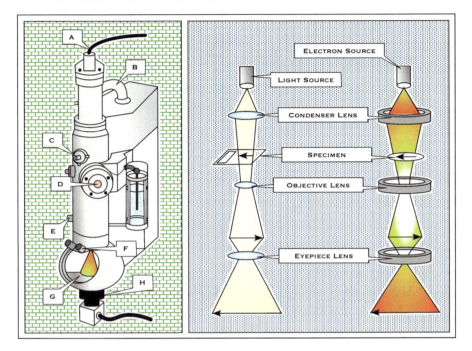

Fig. 8.3 Anatomy of the transmission electron microscope
This figure shows a schematic representation of some of the most important components of the TEM: (**a**) the high-voltage supply needed for electron generation; (**b**) the connections that direct the vacuum pumps, resulting in an environment free of molecules that could interfere with the pathway of the electron beam; (**c**) the apertures that allow for a condensed and aligned beam to be directed at the sample; (**d**) the sample placement area; (**e**) the apertures at the objective lens that is key for rearrangement of the electron beam; (**f**) binoculars, which are essential for magnifying the projected image by 10×, which is useful for fine focusing of the image within the projection chamber (**g**); and (**h**) the recording area
To more effectively understand the fundamentals of microscopy, this figure also illustrates the layout of an inverted white-light microscope. In both microscopes, after light is emitted by an energy source, the condenser lens aligns and focuses the light on the sample. The light is then transmitted through the sample, and the information produced by the light-sample interaction is retrieved by the objective lens. Finally, this information reaches the projection lenses, analogous to ocular lenses. The former transmits information to a microscope screen or a camera, while the latter delivers it directly to the observer

The diagram depicted in Fig. 8.4 (top panel) shows the inner structure of this microscope, from the electron emission site to the image projection area. The organizational structure of the electron gun is the same as the one previously described in Chap. 7; however, there is a key difference: the initial current supplied in a TEM begins at 50,000 electron volts (or 50 keV) and, in all its classic versions, exceeds 100 keV.

As mentioned in the SEM chapter, the filament is a crucial part of the electron gun. It is the structure responsible for electron emission following the direct application of high-energy currents. Inherent to its role, this component is highly susceptible to short or medium-term deterioration or burnout. With the goal of extending its lifespan, filaments have been designed using different materials (Fig. 8.4, bottom panel).

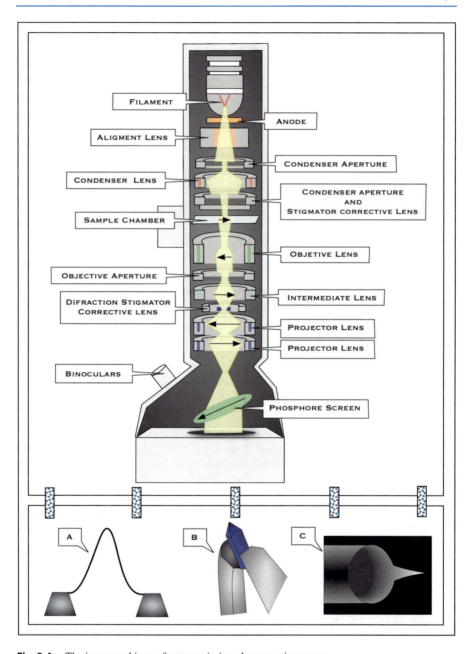

Fig. 8.4 The inner workings of a transmission electron microscope
The top part of this figure shows a diagram of the interior of a TEM, displaying with most detail the area of the electron gun, where the filament is located. Each section of the microscope will be described in greater detail in subsequent images. The bottom panels show the different types of filaments that can be incorporated into the electron gun: (**a**) tungsten, (**b**) lanthanum hexaboride, and (**c**) field emission (which is composed of tungsten and zirconium oxide)

8 The Transmission Electron Microscope

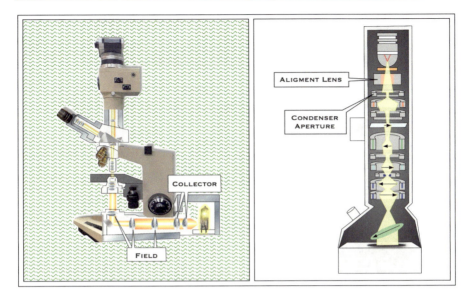

Fig. 8.5 Alignment coils: from chaotic light to an ordered beam
Subsequent to electron emission, one of the primary and initial objectives within the microscope is to uniformly arrange the direction and diameter of the newly formed light beam. In a brightfield microscope (left panel), this important function is performed by collector lenses located right in front of the white-light lamp. The same principle is applied in the TEM (right): once the electron beam exits the electron gun, it is focused and aligned by alignment coils, which ensure the electrons are orderly directed toward the first condenser aperture

Underneath the electron gun (Fig. 7.7, Chap. 7), we can find the alignment coils. To explain the purpose of these electromagnetic lenses in a simple and concise manner, we can compare them to the collector lenses on a white-light microscope (Fig. 8.5).

A TEM is also equipped with condenser apertures, which are adjustable elements that can be matched with the magnification the user chooses. When low

Fig. 8.4 (continued) The filament plays a fundamental role in the TEM, as it is responsible for electron emission. Its characteristics directly influence the quality and resolution of a TEM micrograph. Economically, tungsten filaments (**a**) are the most affordable, capable of withstanding temperatures above 3000 °C. They are used in biological applications for standard electron microscopy, as they are not very efficient at generating narrow-diameter electron beams and thus result in low-light intensity. This also makes them less suitable for analyzing inorganic materials or high-resolution biological samples. On the other hand, lanthanum hexaboride filaments (**b**) are preferred for inorganic analysis and high-resolution studies, as they generate a larger number of electrons and, consequently, a higher light intensity. Due to their high precision and light intensity, they produce an emission diameter ten times smaller than tungsten filaments. They are commonly used in fields such as nanotechnology. However, their acquisition and maintenance costs are higher, but the benefits in terms of resolution and lifespan are noticeable. Regarding light emission diameter and intensity, field emission filaments (**c**) are the most precise and powerful, producing images that stand out for their high resolution and superior signal-to-noise ratio. As expected, they are the most expensive but also the most durable

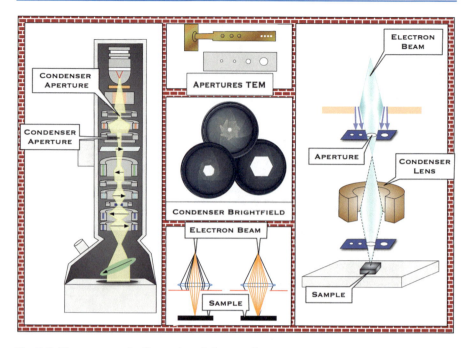

Fig. 8.6 The apertures: the fine-tuning of electron alignment

The primary and secondary apertures of the condenser lens (left) are dynamic elements responsible for selecting the optimal "spot size" needed for interaction with the sample. These apertures eliminate electrons that are scattered or far from the main axis of the light cone, ensuring they arrive uniformly at the condenser lens; therefore, they help to eliminate chaotic illumination by preventing electrons far from the beam's central axis from reaching the sample. Remember, the aperture size is always inversely proportional to magnification. Its setting should be considered for low, medium, and high magnifications

In electron microscopes, there are typically three to four fixed apertures with static diameters (upper middle panel). Functionally, they are comparable to the condenser aperture in a brightfield microscope, although in the latter, the aperture is physically adjustable in the same element—it is a diaphragm (center middle panel). Aperture selection relies on a retractable rod that positions a non-dynamic slit, which is physically controlled from outside the microscope. This mechanical control allows the electron beam to be concentrated on either small or large areas of the sample. Focusing energy on a small sample area is ideal for higher magnifications and high-voltage applications (left, lower left panel). Conversely, selecting larger apertures is suited for low magnifications and low-energy observations (right, lower left panel). Overall, this strategy is well organized to direct electrons toward the sample (right panel)

magnifications are being used, wider apertures covering broader fields are chosen, as they illuminate larger areas of the sample. The opposite is true for higher magnifications, as high energy and smaller illumination areas are needed to focus the electrons to the desired points of the sample (Fig. 8.6).

Directly below the primary apertures, we can find the condenser lens. This lens interacts with the electron beam, focusing and directing it toward the sample. Following this structure, the secondary condenser apertures are located. As mentioned in the corresponding image, these serve to eliminate electrons that have been deflected from the central axis of the electron beam due to diffraction, as these tend

8 The Transmission Electron Microscope

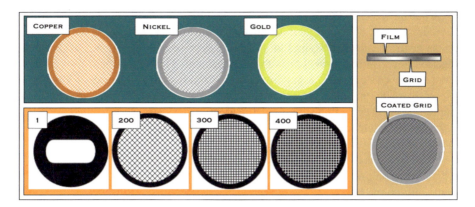

Fig. 8.7 The grids: a window with fine metalwork
Grids are small discs with diameters of 3 mm that can be made from various materials. The most common are copper, nickel, and gold (on top), with nickel and copper grids being the most widely used. Copper grids are ideal for samples that require minimal solution exchange, such as those destined for ultrastructural analysis, due to their high susceptibility to oxidation in humid environments. In contrast, nickel grids are preferred for procedures involving significant buffer exchange, such as the immunogold technique, because they are resistant to oxidation caused by liquid media
The mesh density of grids is essential for sample analysis. Using a grid that has too much mesh density when analyzing tissue samples can obscure data, as the grid's bars may interfere. Conversely, using a low-mesh grid for nanometric samples may lead to sample loss during preparation or instability under the electron beam. The figure illustrates grids with 1200, 300, and 400 meshes (below), which are commonly used for tissues and cells
Although grids provide basic support, they are insufficient for ultrathin and fragile samples (typically 60–90 nm thick) when exposed to hundreds or thousands of volts. Therefore, a coating is required to offer support, adhesion, and stability (right). For high-stability or prolonged observation, grids must be coated with films like formvar, butvar, collodion, parodion, or pyroxylin. Formvar-carbon films offer additional stability under high magnifications and energy levels above 100 keV. While double coating is beneficial, it is not recommended for low magnifications as thick areas may be mistaken for sample content

to generate image aberrations. Thus, these apertures finely regulate the diameter of the electron beam irradiating the sample located below. The sample in question is supported thanks to a TEM grid that in turn is securely kept in place by another specialized holder. These grids are flat discs that can be made from various materials such as copper, nickel, and even gold. The main requisite is biocompatibility with the sample and lack of distortion of the magnetic field of the lens. Due to the chemically inert nature of gold, its use is popular in this context.

Grids can also be characterized by the number of openings they contain. Each opening is referred to as a mesh, and grids are distinguished by the amount of mesh per inch contained (mesh/in). In most cases, grids have 200, 300, or 400 mesh/in. The shape of the mesh varies, with square or hexagonal geometries being the most common; the latter ones result in larger observation areas. How should one choose the appropriate grid based on the sample? The number of apertures should be selected according to the size of the sample; for instance, nanoparticles require a high mesh count to stay stabilized on the grid as the electron beam collides with them. On the other hand, ultrathin sections of large tissue samples can be placed on grids with fewer apertures or even on a single-mesh grid (Fig. 8.7).

A plastic electron-transparent film is placed between the sample and the grid to provide support, adhesion, and adequate stabilization (Fig. 8.7). Its non-electron-dense nature ensures that this film will not interfere with the transmission of the electron beam after interacting with the specimen. This component also protects from continuous exposure to high-energy electron radiation. The use of this film is especially important with grids that have a low mesh count. Without the film covering the grid and providing stability, large areas of the sample could bend upon colliding with the electron beam and might even fracture or burn from the radiation energy in the aperture areas of the grid.

Some samples may be suitable to be placed on the grid without a film, but these will have to be primarily adhered to the grid bars, which are physically electron-dense in nature. However, sample remnants that withstand the radiation usually protrude slightly from the grid bars and thus provide little or no usable information. Thus, to get the most information out of the whole sample, it is often recommended to use a film.

Films can be made from various materials, such as:

(a) Formvar (polyvinyl formaldehyde resin): Recommended for energies above 100 keV
(b) Butvar (polyvinyl butyral): Suitable for negative staining due to its hydrophobic nature
(c) Nitrocellulose-based polymers (such as collodion, parlodion, and pyroxylin): Better suited for energies below 100 keV, and thus for use at medium to low magnifications

Additionally, there are films containing mixed coatings, such as collodion with carbon and formvar with carbon. These make the plastic coatings much more stable.

Regardless of the grid and film material, the sample must be placed in a sample holder for its proper observation. The TEM sample holder at its thinnest end has a small clamping ring to hold the grid by its edges, allowing the sample to be exposed to the electron beam traveling along the central axis of the microscope column (Fig. 8.8).

Once the primary electron beam has interacted with the sample, the same phenomena described in Fig. 7.9 (Chap. 7) typically occur, with the difference that in a TEM, the voltage used is higher. This allows the information from the sample to be acquired with greater precision, both in resolution and in ionic content. The latter is possible when the microscope is equipped with energy dispersive spectroscopy (EDS) (see previous chapter for more information). Regarding the interaction of the electron beam with the sample, two outcomes are particularly relevant for this type of microscopy: absorption and transmission. Both elements give rise to the classic chromatic scale features—comprising black, gray, and white tones—characteristic of TEM micrographs (Fig. 8.9, 8.10, and 8.11).

Located below the sample we would find the objective magnetic lenses, which, as occurs in light microscopes, provide the initial magnification. Once the light-sample information has passed through these lenses, it reaches the diffraction lenses and continues toward the objective aperture. This aperture, along with the

8 The Transmission Electron Microscope

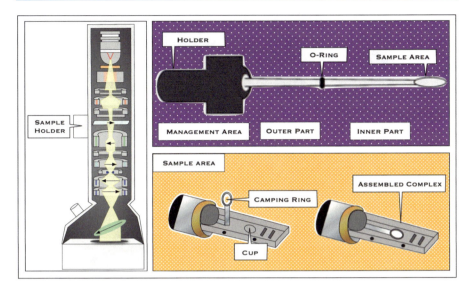

Fig. 8.8 TEM sample holder: a fencing foil for electron microscopists
The sample holder is located at the center of the column (left). It consists of three main parts (right top): (1) the handling area, which is equipped with a crank; (2) the external section, which extends from the end of the crank to the O-ring, which is a polymer spinner washer that acts as a seal to maintain the vacuum inside the sample chamber; and (3) the internal section, which is positioned at the column's core. This section holds the sample in alignment with the electron beam at the top and with the objective lens at the bottom. The internal segment (right below) includes a "cup" with a diameter slightly larger than 3 mm, where the grid containing the sample is placed. To immobilize the grid, its edges are mechanically secured by a clamping ring, keeping the grid-sample assembly stable and ready to receive high-energy electron collisions. Materials used for the cup and clamping ring must be minimally reactive to electrons to prevent X-ray fluorescence. Commonly, beryllium is the material of choice. Sample holders can accommodate various numbers of cups, ranging from single to multiple configurations. Cups also vary in their angles of inclination, which is useful when performing diffraction imaging. Some of them even offer dynamic tilt and rotation capabilities. Because of its distinctive shape, the TEM sample holder is often compared to a fencing foil

diffraction lenses, helps focus the electron beam onto a corrective lens, which eliminates aberrations from the information before it interacts with the projection lenses. The projection lenses will subsequently direct the information to the image projection chamber for its observation and recording (Fig. 8.12).

In the projection chamber, the light, which contains the information from the irradiated sample area, is converted into a photonic image by a phosphor screen. This tool acts as an electron-to-photon transducer and generates an image composed of thousands of voltage-dependent photons. This image will be of a fluorescent green color due to electron–phosphor interactions (Fig. 8.12).

As opposed to the relatively simple image acquisition, sample processing is a complex procedure that often requires patience and long training periods. Tissue samples are often obtained through the process of cardiac perfusion fixation. Diverse fixative solutions are used, such as those containing a mixture of two aldehydes: paraformaldehyde and glutaraldehyde (well-known as Karnovsky's fixative). If the

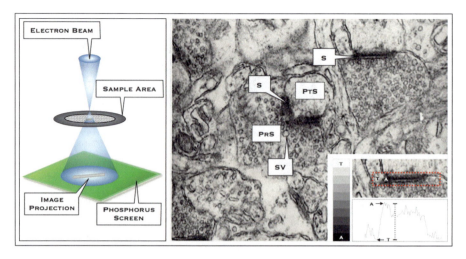

Fig. 8.9 Synaptic junctions under the gaze of a TEM
The schematic representation observed at the left of this figure shows an electron beam coming from the condenser lens colliding with a sample. When the electron beam interacts with the sample, it produces various phenomena, with absorption and transmission being the most significant ones. Absorbed electrons generate dark tones, while transmitted electrons create bright tones. Due to absorption–transmission phenomena, images such as the photomicrograph pictured in this figure can be created. Here, numerous elements of the neuropil in the central nervous system can be observed. The neuropil is a network of structures that connect one neuron to another or one neuronal nucleus to another. At the center and upper-right region of the image, two highly electron-dense (black) structures, known as synapses (S), are visible. Synapses facilitate neuron-to-neuron communication enabling the transmission of a chain reaction called a nerve impulse. An impulse transports information between regions, either as incoming sensory data or outgoing responses to stimuli. A synapse will include a presynaptic terminal (PrS), which is characterized by the presence of numerous synaptic vesicles (sv) containing neurotransmitters. Upon fusion with the presynaptic membrane, these vesicles release their contents into the synaptic cleft. The neurotransmitter then becomes available for recognition by receptors embedded in the postsynaptic terminal (PtS), which lacks synaptic vesicles
Bottom: This series of monochromatic images illustrates the phenomena of absorption (A) and transmission (T) in a synaptic terminal. The synaptic region (red dotted area) is highly dense due to the presence of neurotransmitters and ions. Here, electrons are primarily absorbed (dark areas). Flanking the synapse are regions where electrons are easily transmitted (bright areas), giving the transmission electron microscope its name. Shades between black and white (gray tones) represent areas of the sample that partially absorb the electron beam to varying degrees. This micrograph was analyzed using a density diagram from the FIJI software, which highlighted zones of high absorption (A) and high transmission (T) to create a gradient (black dotted line) as represented in grayscale

sample is intended for protein localization, which would require a technique identified as immunogold labeling, it is then post-fixed by immersion in the same fixative solution. For cellular ultrastructure visualization, the use of osmium tetroxide as a fixative is necessary to enhance the contrast of membrane lipids. When using cultured cells or non-perfused tissues, the initial step is immersion fixation.

Subsequently, the area to be worked with (which must be around 1 mm^3) will be selected and carefully handled. When using cells, these have to be compacted using a centrifuge. During this step, water content begins to be removed, allowing for its

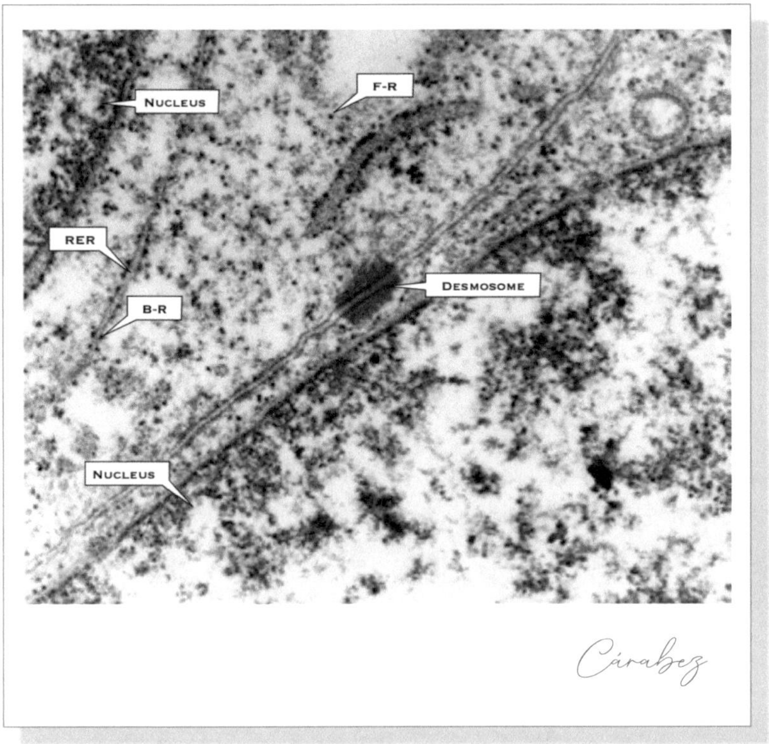

8 The Transmission Electron Microscope

Fig. 8.10 The neuronal kiss
This image shows the nuclei of two cells—one located in the upper-left corner and the other in the lower-right corner. Visible structures include distinguishable by bound ribosomes (B-R) cisternae of the rough endoplasmic reticulum (RER), and free ribosomes (F-R). Most notably, at the center of the image, a density appears on either side of the plasma membranes of the two neurons. This structure is identified as a desmosome, a site of ion exchange between neurons, emphasizing their origin from embryonic neuroepithelium

replacement with a resin that is liquid at room temperature but solid (plastic) at temperatures above 60 °C. Having a sample with sufficient hardness allows for ultrathin sectioning and for withstanding the high temperatures generated by the electron beam. Replacing water with plastic resin also helps to keep the interior of the column free of water molecules. This ensures optimal electron beam movement and minimal contamination. A graphical summary of sample processing can be seen in Fig. 8.13.

This environment mentioned above which is required for the free movement of electrons and that must be free of any type of particles (such as water) is a vacuum. When compared with an SEM, the TEM is integrated with more vacuum pumps. The former generally has two, while the TEM is equipped with at least four, which are known as: (1) rotary, (2) oil diffusion, (3) turbomolecular, and (4) ion getter pumps.

Fig. 8.11 The muscular tapestry: A wondrous woven magic
The title of this figure refers to a high-magnification image of a segment of voluntary skeletal muscle tissue, also known as a red muscle due to its high content of myoglobin, which is a protein similar to the hemoglobin in red blood cells that gives it a reddish hue. Muscles contract or stretch based on instructions received through neuromuscular junctions (N-mj), distinguishable by the presence of synaptic vesicles. Perfectly pictured here is the defining feature of this muscle type: the highly organized structure essential for its contraction–relaxation function

8.1 When Should a Transmission Electron Microscope (TEM) Be Used?

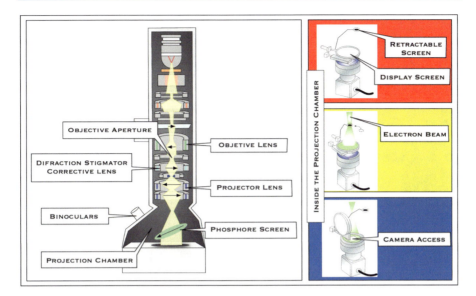

Fig. 8.12 From objective lens to image projection: the last journey of electrons
Left panel: The process of magnification begins with the objective lenses and is highly dependent on the manufacturer of the microscope. The outgoing information from these electromagnetic lenses is corrected using the objective apertures. Like any beam of light, an electron beam expands its diameter (or cone) after being focused. Thus, placing an aperture beyond the objective lenses to refine the central axis of the light cone is an effective strategy. However, as the electron beam moves through the TEM column, it experiences distortions in its wave, particularly at high magnifications. This makes a correction lens essential to fix any astigmatism and diffraction patterns and sends the corrected beam to the projector lenses, which precede the projection chamber
Right panels: The projection chamber features an external binocular system that can magnify the image ten times around. This binocular system is used for fine focusing and selecting the center of the image for acquisition. When using the binoculars, a small retractable circular screen is employed. Below this small screen lies the larger viewing screen, also circular and significantly larger in diameter. Both screens are coated with phosphor, which converts electron signals into green photons upon collision. The higher the energy applied to the sample, the greater the photon generation, resulting in higher resolution and a better signal-to-noise ratio. The generated image can be viewed inside the chamber or projected to a computational system via a high-resolution camera

Transmission electron microscopy has been enhanced in previous years by incorporating the ability to generate three-dimensional images through micrographs obtained from serial sections that can be reconstructed to create 3D models. This software-based technique now allows users to observe the complete subcellular morphology of virtually any biological specimen. Additionally, the TEM has recently incorporated the use of samples treated at ultra-cold temperatures (a technique known as cryo-electron microscopy), which is useful to reveal the structure of proteins and other molecules with much greater accuracy, and using smaller sample quantities, than traditional techniques like spectroscopy.

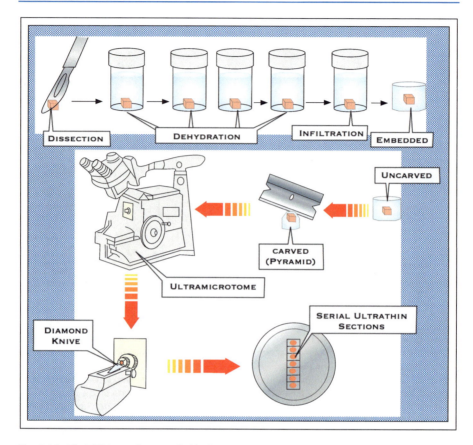

Fig. 8.13 The TEM sample: not suitable for eager hands or excessive self-confidence

As it has been previously mentioned, the transmission electron microscope operates based on the same technical principles as a light microscope. Thus, it requires a short period of training to understand its basic operation. However, sample preparation is a distinct discipline that could be compared to the intricate craftsmanship of jewelry. Usually, this aspect of TEM work often takes years of practice before proper preparation of specimens can be achieved

As illustrated in the diagram, after fixing a tissue sample of approximately 1 mm3 in size should be selected; then, the dehydration process begins. This involves using an alcohol gradient that starts with low concentrations and ends with a couple of absolute ethanol baths. At this point, the tissue is free of water and ready to receive a 1:1 resin-absolute ethanol mixture, this will facilitate absolute resin penetration in its liquid state, this step is known as infiltration

Following infiltration, the tissue is exposed to pure resin to remove any ethanol remnants. This step, known as embedding, takes several hours at room temperature, followed by heating to 58–60 °C for 20 h around. At these temperatures, the resin polymerizes, providing sufficient hardness to allow the cutting of ultrathin sections between 60 and 90 nm in thickness. This thinness ensures that electrons can pass through the sample

The sections are made using an instrument called an ultramicrotome, which is equipped with a stereoscopic optical viewing system, a cutting area, and a crank mechanism with precise controls for systematic sectioning. The knives used are typically made of glass or, due to its superior hardness, diamond. Once the sections are obtained, they are mounted onto grids and stained with heavy metal solutions to create a contrast that is suitable for observation

8.1 When Should a Transmission Electron Microscope (TEM) Be Used?

TEM provides the highest resolution currently available, allowing for the visualization of structures as small as ribosomes (20–30 nm) and chromatin granules (as small as 10 nm in diameter). The conventional and most widely used TEM imaging mode involves transmitted electron beams passing through the sample to reveal its ultrastructure. However, like in brightfield microscopy, TEM can also produce darkfield micrographs using a technique called *negative staining*. This method captures diffracted electron beams to form an image, and while it has a lower resolution than conventional TEM, it provides useful information about a specimen's contours.

A key technique for localizing molecules at the subcellular level is immunogold labeling, where primary or secondary antibodies are conjugated to colloidal gold nanoparticles of various sizes, ranging from just over 1 nm to around 40 nm. For molecules located deep within the cell, such as in the nucleus, smaller gold particles are preferred to ensure penetration. Conversely, larger gold particles are ideal for labeling external regions, such as the plasma membrane.

TEM can also visualize enzyme-based stains commonly used in brightfield microscopy. Some histological stains, such as *horseradish peroxidase* (HRP) and *hematoxylin*, appear to be electron-dense (darker) under TEM. Similarly, certain fluorochromes like *quantum dots* exhibit electron density, facilitating correlative microscopy techniques that combine fluorescence and electron microscopy. Additionally, serial ultrathin sections obtained with an ultramicrotome can be imaged sequentially to create high-resolution 3D reconstructions, similar to confocal microscopy. However, TEM-based 3D reconstructions require specialized software and can generate detailed models of entire organelles.

Another major TEM technique is *cryo-electron microscopy* (cryo-EM), which is the method of choice for the structural studies of biomolecules at the sub-nanometer level. Cryo-EM enables near-atomic resolution imaging of proteins, viruses, and macromolecular complexes in their native state, making it a powerful tool in structural biology.

Electron microscopy, just like the other forms of conventional microscopy that were introduced in this book, is unquestionably a tool that allows for the continuous redefining of boundaries within current scientific knowledge. Many innovators, inventors, and creators of microscopes and techniques associated with microscopy have been awarded all types of prestigious accolades and are considered authentic pioneers of science. The creativity and expertise that is often required when operating a microscope to its full capability frequently result in photomicrographs that transcend science and enter the realm of art. Much like the Chinese ink illustrations and photomicrographs featured in this book, both elements provide great harmony to the academic words and illustrations. The human curiosity that galvanizes us to observe what is invisible to the naked eye knows no bounds. Thanks to the advent of super-resolution microscopy, the golden age of *in vivo* biological observations in high resolution has just begun. We are now equipped to observe in real time the unpredictable phenomena and exquisite biological forms that had been kept hidden from our sight and understanding. This is also what we call: "Microscopic Wonders: The Science of Seeing the Invisible."

Further Reading

For electron microscopy basis: Bozzola & Russell (1999), Dykstra & Reuss (2011), Hall (1953), Hall (1985), Harris (1991), Egerton (2006), Merchant (1994), Nguyen & Harbison (2017), Yacaman & Reyes (1995), Vázquez-Nin & Echeverría (2000), Williams & Carter (1997).

For Electron microscopy samples: Ayache et al. (2010), Sasaki et al. (2022).

A text book illustrated of the ultrastructure of the nervous system: Peters et al. (1991).

References

Ayache J., Beaunier L., Boumendil J., Ehret G., Laub D. 2010. Sample preparation handbook for transmission electron microscopy: techniques. Springer New York. XXV, 338

Bozzola JJ, Russell LD (1999) Electron microscopy: principles and techniques for biologists. Jones and Bartlett Series in Biology

Dykstra MJ, Reuss LE (2011) Biological electron microscopy: theory, techniques, and troubleshooting. Springer, New York. https://doi.org/10.1007/978-1-4419-9244-4

Egerton RF (2006) Physical principles of electron microscopy: an introduction to TEM, SEM, and AEM, vol XII. Springer, New York, p 202

Hall CE (1953) Introduction to electron microscopy. McGraw-Hill, Hightstown, p 451

Hall CE (1985) The beginnings of electron microscopy, advances in electronics and electron. Physics Academic Press, New York

Harris JR (1991) Electron microscopy in biology. A practical approach. IRL Press/Oxford Univesity Press, Oxford, p 328

Merchant H (1994) El inicio de la microscopía electrónica en México. Ciencia y Desarrollo

Nguyen JNT, Harbison AM (2017) Scanning electron microscopy sample preparation and imaging. In: Espina V (ed) Molecular profiling. Methods in molecular biology, vol 1606. Humana Press, New York. https://doi.org/10.1007/978-1-4939-6990-6_5

Peters A, Sanford LP, Henry de FW (1991) The fine structure of the nervous system: neurons and their supporting cells, 3rd edn. Oxford University Press, Oxford

Sasaki H, Arai H, Kikuchi E, Saito H, Seki K, Matsui T (2022) Novel electron microscopic staining method using traditional dye, hematoxylin. Sci Rep 12:7756. https://doi.org/10.1038/s41598-022-11523-y

Vázquez-Nin G, Echeverría O (2000) Introducción a la Microscopía Electrónica Aplicada a las Ciencias Biológicas. Facultad de Ciencias, UNAM. Fondo de Cultura Económica, p 163

Williams DB, Carter CB (1997) Transmission electron microscopy: a textbook for materials science. Micron 28(1):75–75

Yacaman MJ, Reyes J (1995) Microscopía Electrónica: Una visión del microcosmos. Consejo Nacional de Ciencia y Tecnología. Fondo de Cultura Económica, p 143

Glossary

Aberration correction The elimination of image distortions through the use of multiple lenses or through the placement of fixed or dynamic apertures that reduce or block peripheral light rays (the primary cause of aberrations).

Aberration A distortion or blurring of an image caused by the interaction of light with the design and shape of the lenses.

Absorption of light The retention of light energy by elements of a sample, also referred to as optical absorption.

Airy disk An optical diffraction phenomenon that occurs when light passes through a circular tiny aperture. The interaction between light and the aperture produces an image plane with a concentric ring pattern of alternating light and dark bands.

Airy unit The central diffraction point (central disk) in the image plane within the Airy disk.

Amplification In microscopy, the digital enlargement of a specimen. Higher amplification results in lower resolution since it does not involve the use of lenses.

Anode A component of the electron gun and photomultiplier that attracts negatively charged particles due to its positive nature.

Arc lamp A lamp with slightly separated positive and negative electrodes, typically made of tubular quartz glass with a central bulb. These lamps commonly contain mercury or xenon gases and are high-energy light sources.

Astigmatism An imaging defect where different focal planes form along the axial axis due to lens shape and inherent deformations. The distance between focal points in this aberration is known as the astigmatic difference.

Auger electron An electron ejected from an atom due to displacement caused by the collision of an inner electron that has absorbed high-energy radiation.

Autofluorescence Also known as primary fluorescence, it is the ability of certain organic and inorganic specimens to emit fluorescence in response to exposure to specific wavelengths of radiation.

Axial resolution The minimum distance between two points required for them to be observed as separate in the Z-axis (thickness or 3D plane) of a sample.

Backscattered electrons Electrons that reflect back toward the source after interacting with the sample surface. The intensity of these electrons depends on

© The Editor(s) (if applicable) and The Author(s), under exclusive license to Springer Nature Switzerland AG 2025
A. Rosas-Arellano et al., *Microscopic Wonders*,
https://doi.org/10.1007/978-3-031-92559-7

and is proportional to the atomic number of the sample's atoms. Therefore, elements with higher atomic numbers generate a greater intensity of backscattered electrons.

Beam of light A directed stream of light rays emitted from a light source.

Bias In electron microscopy, a controllable resistance that adjusts the voltage difference between the anode and cathode. A greater voltage difference attracts more electrons.

Binocular microscope A microscope equipped with two eyepieces, designed to enhance the field of view and provide a more comfortable observation experience.

Brightfield microscope A microscope that uses white light as its illumination source. It is also known as a transmitted light microscope or compound microscope.

Cathode In electron microscopy, the component of the electron gun that emits negatively charged particles. Cathode rays are emitted from the filament after being exposed to high-voltage radiation.

Chimeric proteins Also known as fusion proteins, these molecules are composed of at least two genes that, when translated within a cell, form a single molecule.

Chromatic aberration A color distortion observed at the edges of sample components when viewed under a microscope. It is caused by focal displacement along the optical axis due to the different wavelengths of white light and electron-generated light.

Chromophore The segment of a fluorochrome responsible for giving color to fluorescent light emission.

Coherent light Light composed of waves with the same wavelength, phase, and frequency—meaning the waves are monochromatic and their peaks and troughs align perfectly.

Collector lens A set of lenses strategically placed inside a microscope to gather emitted light from a source, align it, and redirect it to other lenses or mirrors, depending on the microscope's design.

Coma aberration An off-axis displacement of light, also known as "off-axial aberration." It occurs when light refracts through lenses at increasing angles of incidence or due to improper alignment of lenses or light sources. This aberration creates asymmetrical images that gradually increase in size, resembling the shape of a comet, hence the term "coma."

Compound microscope A microscope consisting of two or more lenses, also referred to as a brightfield microscope.

Condenser lens Also called a positive diopter lens or simply a positive lens, it is used to correct light divergence and align it into a directed beam.

Confocal microscope A high-contrast microscope that employs a spatial diaphragm to eliminate out-of-focus light by selecting a single focal plane. Combining two or more focal planes allows for the reconstruction of 3D images.

Continuous-wave laser A lighting source that emits a continuous laser beam during its activation period.

Convergence of light Parallel light rays that, upon passing through a positive lens, converge at a single point along the lens's axial axis.

Coverslip A thin, translucent sheet placed over a sample to be observed under a microscope. The standard thickness is 1.7 mm, and its function is to provide a flat observation surface while protecting the sample from contaminants or dehydration.

Cryo-electron microscopy A method used to determine the structure of organic and inorganic molecules using a transmission electron microscope with a sample chamber at ultra-low temperature ($-160\,°C$ around). Unlike crystallography, this technique requires a smaller sample quantity.

Curvature of field aberration Also known as Petzval curvature, this aberration appears as an out-of-focus curvature at the edges of an image projection, with a focused area along the axial axis. It partially disrupts sample focus, as adjusting the focus at the edges results in loss of focus at the axial region.

Dark-field microscopy A variation of brightfield microscopy primarily used for observing live samples. In this method, the condenser blocks axial light beams, allowing only peripheral light to pass through. The angled illumination creates bright edges around structures against a dark background.

Dichroic mirror An optical component capable of reflecting a specific wavelength while allowing a different wavelength to pass through.

Differential interference contrast (DIC) microscopy Also known as Nomarski microscopy, DIC is a technique applied in brightfield microscopy to provide high contrast for unstained live or fixed samples. A polarizer first converts polychromatic light into monochromatic light, which then passes through a prism to split it into orthogonal beams. These beams interact with the sample's edges, and upon exiting, they pass through a second prism that realigns them along the axial axis. A final polarizing filter (analyzer) is used before observation. DIC microscopy enhances contrast at internal and external structural edges without staining.

Diffraction The deviation of light waves from their original path due to interaction with a physical object or the edges of a slit equal to or smaller than the wavelength in question.

Dispersion of light The separation of white light into its component wavelengths.

Distortion A monochromatic aberration that alters image shape.

Divergency of light The separation of light beams from the axial axis. Divergent lenses are named for redirecting central light rays toward peripheral regions.

Dynode A positively charged component within a photomultiplier tube that attracts negatively charged particles such as photoelectrons and electrons. The charge of dynodes can be adjusted to control the number of particles attracted.

Electromagnetic lens Analogous to physical lenses, electromagnetic lenses focus or spread electron beams. They consist of a cylindrical copper wire coil (solenoid), a pole piece, and an iron casing that generates a magnetic field to attract and align the electron beam.

Electron gun Also known as an electron emitter, it generates high-energy thermal electrons. It consists of a high-voltage emission source, a filament (cathode), a Wehnelt cylinder (shield), and an anode.

Electron A fundamental subatomic particle with a negative charge, currently considered indivisible.

Emission filter   A polarizer designed to interact with the fluorescence emitted by a sample, eliminating wavelengths that do not correspond to the desired range. This allows only a specific range of wavelengths to pass through.

Energy dispersive X-ray spectroscopy (EDS)   A microanalytical technique used to determine pixel by pixel the elemental composition and concentration in organic and inorganic samples.

Excitation filter   A polarizer that selects a specific wavelength range from a polychromatic light source, typically white light.

Eyepiece lens   A component of a microscope's optical system, consisting of a single lens or a set of lenses. Its function is to provide the final magnification of the sample when viewed through a camera or the human eye.

Field lenses   Converging lenses located between the collector lenses and the condenser. Their function is to keep the light beam aligned. Some modern objectives contain these lenses in their inner structure, where, in addition to aligning the light, they also act as relay lenses to provide magnification to the sample.

Filament   Also called the cathode in an electron microscope, it is an element of the electron gun that emits electrons when receiving high-voltage thermal energy. One of the most commonly used materials for filaments is tungsten.

Fluorescence loss in photobleaching (FLIP)   A microscopy technique used to study the cellular or intracellular dynamics of molecules of interest and their mobility between two defined regions while considering time and space in relation to the fluorescence lost as a consequence of continuous radiation.

Fluorescence recovery after photobleaching (FRAP)   A technique for temporally analyzing the mobility of fluorochrome-labeled molecules within living cells or tissues. This technique involves fluorescence loss in a defined region after a brief exposure to high-intensity radiation. If fluorescence recovers, it suggests there are interactions between the first region and a second defined region within the sample.

Fluorescence resonance energy transfer (FRET)   A technique used to determine the interaction between two molecules within tissues or living cells. If the fluorescence emission of one molecule of interest can activate the emission of a second molecule, it is considered a positive interaction. The maximum distance between the two molecules for this to occur is 10 nm so this colocalization can molecule–molecule interaction.

Fluorescence   The property of certain molecules to emit luminescence at a specific wavelength in response to energy absorption.

Fluorochrome   A molecular complex that absorbs energy and subsequently emits light through its functional groups, known as fluorophores.

Fluorophore   The functional group of a fluorochrome responsible for fluorescence emission at a wavelength determined by the chromophore.

Focal distance   Also known as focal length, it refers to the distance between the optical node of the lens and the focal node. Changes in the positive or negative nature of the lenses directly alter the focal distance.

Focal plane   Also referred to as the focal point or focal node, it is the region where incident light rays converge to form an image.

Galvanometer mirror   These mirrors, of which there are usually two in a laser scanning microscope, are ultra-fast elements with movement orchestrated by galvanometer motors. The mirrors steer the laser scan over the sample in the x–y planes (one mirror per axis).

Immersion media   Liquid media with defined densitometric characteristics higher than those of air, used to increase the resolution power of a microscope. The increased density allows for greater recovery of light information traveling from the upper edge of the sample to the objective's front lens. The most common immersion medium is oil, which can be made from natural or synthetic materials and has a refractive index similar to that of a microscope's objective lenses.

Immunogold labeling   A technique used in electron microscopy to determine the presence of antigens in a sample. The antibodies used in this method are tagged with colloidal gold particles, which appear as electron-dense circles in transmission electron microscopy. This technique allows for a highly precise localization of a molecule at the subcellular level.

Interface   The separation or boundary between two material media with clearly distinct densities or compositions.

Inverted microscope   A type of microscope where the light source and the condenser are located at the top, while the objectives are positioned in the middle-lower section, with the front objective lens facing upwards. This type of microscope is highly useful for observing live samples positioned on the lower surface of a preparation, such as culture dishes.

Iris diaphragm   An adjustable diaphragm made of thin, electron-dense plates, located above the light source. Its mechanism for opening and closing to regulate light passage resembles the iris of the human eye and other species, hence its name.

Köhler illumination   A method that provides optimal light alignment in a brightfield microscope. This technique is used to maintain uniform illumination across the sample and achieve high-quality image acquisition. Köhler adjustment is performed in out-of-focus planes relative to the sample to prevent its components from affecting the final image interpretation.

Laser   Acronym for *light amplification by the stimulated emission of radiation*. Coherent, monochromatic, high-energy, and collimated light. Some types are made using gas, liquid, or solid-state materials.

Lateral resolution   The minimum distance that is required between two points for them to be observed independently in the x–y axes (2D plane) of the sample.

Lens   An optical element that magnifies an object. Depending on its shape, it can concentrate or disperse light. Lenses are made of translucent materials; however, harder materials like glass or crystal generate more uniform light pathways. When light interacts with lenses, phenomena such as reflection, refraction, dispersion, interference, and resonance occur.

Magnification   In microscopy, it refers to the enlargement of a specimen using lenses, within the limits determined by their inherent characteristics and the illumination energy employed. Higher magnification generally implies greater resolution.

Medium The microenvironment through which light travels to make sample observation possible using a microscope. The most common media are air, water, and oil. In microscopy, "medium" is represented by the letter "n."

Microscope An instrument that enables the observation of objects and details, not visible to the naked eye, through the single lens (simple) combination of two or more lenses (compound).

Microscopic An adjective referring to anything related to a microscope. It is also used to describe samples that are too small to be seen by the human eye.

Microscopy A set of techniques and methods associated with an instrument (microscope) that allows for the observation of living or fixed specimens beyond the range of human vision.

Monocular microscope A microscope equipped with a single eyepiece. This design is considered obsolete due to lack of comfort and limited field of view.

Negative diopter lenses Also known as "divergent lenses" because they separate light from its axial path (optical axis). The diopter values become increasingly negative as the curvature of the lens approaches the tangential axis relative to the optical axis.

Numerical aperture A dimensionless geometric parameter (meaning it is not a physical object) that indicates the angle of light required to be collected by a lens. For example, only a portion of the white light transmitted through a sample will be captured by the objective lens of a brightfield microscope. This portion of light entering the objective is represented by a cone with the vertex pointing toward the sample.

Objective lens An optical component of a microscope that provides the initial magnification of the observed sample. These lenses vary in design and magnification and may include corrections for optical aberrations or attachments for DIC, TIRF, phase contrast, fluorescence, and other microscopy techniques.

Phase contrast (Ph) A microscopy technique that uses paraxial light and eliminates central illumination to enhance the edges of a translucent or colorless sample.

Photobleaching Also known as "fading," it is the loss of fluorescence emission capability due to prolonged exposure to radiation at a specific wavelength.

Photocathode A component of photomultiplier tubes that converts photons into photoelectrons, depending on the incident light (i.e., the number of photons received).

Photolithography coat A coating that provides photoresistance to a microscope's optical system. It is commonly used in widefield, confocal, single-photon, and two-photon microscopes due to the high-energy radiation exposure of the optical system.

Photomultiplier tube (PMT) A highly sensitive light detector capable of detecting very weak light signals, converting them into electrical signals, and subsequently amplifying them for visualization using a digital system.

Photon A fundamental particle of light. Since there is no evidence of subcomponents or substructures, it is defined as an elementary particle. It has no charge or rest mass, and its speed corresponds to the speed of light in a vacuum.

Glossary 161

Photostability The ability of a fluorochrome to receive continuous radiation exposure and consistently emit a specific amount of photons.

Pinhole A dynamic and automated aperture that functions like a camera diaphragm or a microscope condenser. This small opening eliminates out-of-focus information to produce sharp, high-contrast images.

Positive diopter lenses Also known as convergent lenses because they bring light together along their axial path (optical axis). The diopter values become more positive as the curvature of the lens moves further from the tangential axis relative to the optical axis.

Primary electron beam A high-energy electron beam that emerges from the electron gun and travels through electromagnetic lenses toward the sample.

Projector lens The final magnifying lens of a transmission electron microscope, responsible for enlarging and projecting the image onto the phosphor screen, hence the name "projector lens."

Pulsed tunable laser A laser that alternates between on and off states. Its wavelength can be adjusted to generate a specific wavelength.

Quenching A process that reduces the fluorescence intensity of a fluorochrome due to continuous radiation exposure. Quenching does not imply the complete loss of fluorescence.

Reflection of light The bouncing of light beams off a generally smooth and shiny surface. Every material, regardless of its brightness or smoothness, has a critical angle of reflection. This angle determines when light achieves total internal reflection without being transmitted or refracted from one material to another.

Refraction of light A change in the angle of light as it moves from one medium to another. This change can involve a decrease in the energy and speed of the light beams.

Refractive index A value that indicates how much light changes its angle when transitioning from one medium to another.

Resolution limit A numerical parameter indicating the minimum distance between two objects that allows them to be observed as separate in an optical instrument.

Resolution The ability of an optical system to distinguish two objects as separate.

Scan coil A fast-moving, high-precision electromagnetic lens located above the objective lenses in a scanning electron microscope. Its function is to provide magnification and control the deflection of the electron beam to perform point-by-point scanning of the sample.

Scanning electron microscope (SEM) A microscope equipped with an electron source to obtain topographic information about a sample. If equipped with an EDS detector, it can also provide chemical composition data of the sample.

Scattering of light The deviation of light from its original path upon encountering another medium. This deviation depends on the angle of collision and the physicochemical properties of the object.

Scintillator A component of the electron detector in a scanning electron microscope. Its function is to convert electrons from the sample into visible light signals for subsequent image formation.

Second harmonic generation (SHG) An optical phenomenon in which two photons with the same wavelength interact with a nonlinear material (such as crystals or collagen) and give rise to a new photon with twice the energy and half the wavelength.

Secondary electron beam A type of low-energy electron beam generated by the collision of the primary beam with the sample surface.

Secondary fluorescence Fluorescence emission from living or inert specimens that have been introduced to a fluorochrome. These fluorochromes absorb energy and emit fluorescence at known wavelengths.

Simple microscope A single-lens optical device used to magnify specimens. A magnifying glass is an example of a simple microscope.

Slide A thin glass sheet used to support samples for analysis under a light microscope. Slides come in various sizes and shapes, and some have adhesive coatings to secure samples during experimental processes preceding observation.

Spherical aberration A change in focal points within the image plane due to light interacting with a lens. Lenses with curvatures that extend farther from their axial path (greater positivity) are more likely to exhibit this type of aberration.

Spot size The diameter of the electron beam. Larger spot sizes are suitable for lower magnifications and vice versa. The beam diameter depends on the supplied current—higher current combined with a small spot size improves resolution.

Stereo microscope A microscope that provides three-dimensional vision and a long working distance, making it ideal for dissections or observing whole organs and animals. However, its resolution is lower than that of a brightfield microscope.

Synchrotron A type of particle accelerator used to study both the structure and the chemical composition pixel by pixel of organic and inorganic samples. As of 2025, there are approximately 70 synchrotrons worldwide, with the United States having the most, followed by Germany, the vast majority of countries do not have one. Each synchrotron offers different resolution levels, but all achieve subcellular resolutions in the nanometer range.

TEM grids Equivalent to a slide in light microscopy. It is a small metallic disk (e.g., nickel, copper, or silver) about 3 mm in diameter with holes of various shapes. This component holds the sample for observation under a transmission electron microscope.

Thermionic ionization The generation of ions through the application of thermal energy (heat) to a specific material.

Total internal reflection fluorescence (TIRF) A fluorescence microscopy technique that records by light reflection the surface events of a sample within a range of approximately 200 nm.

Transmission electron microscope (TEM) A high-resolution microscope that transmits an electron beam through an ultrathin sample to generate images. TEMs operate in high vacuum, at high energy, and use electromagnetic lenses instead of physical lenses. Techniques such as cryo-electron microscopy and EDS can be coupled with this type of microscope.

Transmission of light A physical phenomenon in which light passes through an object or specimen due to its transparency, translucency, or thinness.

Two-photon microscopy A fluorescence imaging technique that uses a pulsed and tuneable laser for observing fixed or living specimens. This microscope can capture images of thick samples and even living animals at depths of hundreds of microns. The excitation energy required for fluorochromes is typically half of that used in single-photon microscopy, extending the lifespan of fluorochromes.

Upright microscope A conventional light microscope configuration, with eyepieces at the top, objectives below them, followed by the stage and condenser. This setup is ideal for observing fixed samples.

Wavelength of light (λ) The distance between two consecutive peaks or troughs of a wave. Wavelength values decrease from the infrared to the ultraviolet spectrum. Is represented by λ.

Wehnelt cylinder A component of the electron gun, also called the "Wehnelt cap." It serves three important functions: (1) acting as an electrostatic lens to attract and focus electrons, (2) allowing the organized exit of the electron beam through a central aperture, and (3) dissipating thermal energy from the electron gun via its lateral openings.

Widefield microscope A microscope that captures broad fluorescence emission fields along the axial plane of a sample. It is typically coupled with a short-arc lamp as its energy source.

Working distance The distance between the topmost layer of a sample (e.g., the outer surface of a coverslip for fixed samples) and the front lens of the objective.